TIME AND PSYCHOLOGICAL EXPLANATION

SUNY Series
Alternatives in Psychology

Michael A. Wallach, Editor

TIME AND PSYCHOLOGICAL EXPLANATION

Brent D. Slife

STATE UNIVERSITY OF NEW YORK PRESS

Production by Ruth Fisher
Marketing by Lynne Lekakis

Published by
State University of New York Press, Albany

For information, address the State University of New York Press,
State University Plaza, Albany, NY 12246

Library of Congress Cataloging-in-Publication Data
Slife, Brent D.
    Time and psychological explanation / Brent D. Slife.
        p.   cm. — (SUNY series, alternatives in psychology)
    Includes bibliographical references and index.
    ISBN 0-7914-1469-8 (cloth : alk. paper). — ISBN 0-7914-1470-1
(pbk. : alk. paper)
    1. Time—Psychological aspects.   2. Time perception.   I. Title.
II. Series.
BF468.S56   1993
153.7'53—dc20
                                                                    92-21062
                                                                    CIP

10 9 8 7 6 5 4 3 2 1

To Karen
With all my love

Concepts which have proved useful for ordering things easily assume so great an authority over us that we forget their terrestrial origin and accept them as unalterable facts. They then become labeled as "conceptual necessities," "a priori situations," etc. The road of scientific progress is frequently blocked for long periods by such errors. It is therefore not just an idle game to exercise our ability to analyze familiar concepts, and to demonstrate the conditions under which their justification and usefulness depend.

—Albert Einstein

CONTENTS

PREFACE

Time—no topic could be a greater mystery and yet have so profound an influence on all manner of studies, especially the sciences. If a scientist is concerned at all about the organization and measurement of events, then he or she must have a conception of time. Of course, "hard" scientists have long known about this. From Isaac Newton to Albert Einstein and Stephen Hawking, physical scientists have extensively discussed their own particular conception of time and its significance to the discipline.

Interestingly, "soft" scientists have rarely mentioned temporal conceptions. Be assured that this is not because soft scientists do not have such conceptions. Psychology—the topic of the present volume—is just one of many disciplines to use past, present, and future constantly in its explanations and applications. Moreover, all the crucial aspects of its explanation—causation, change, process, behavior, etc.—are highly related to time. It is extremely rare, however, to find *any* systematic discussion of time in the field. Indeed, it is rare to find a psychologist who knows that he or she *has* a conception of time. Time is considered either an aspect of reality or an indisputable axiom, with neither requiring study or debate. This is a mistake. It is a mistake not only because there are different conceptions of time, but also because these conceptions have dramatically differing implications for theory and therapy.

To correct this mistake, we must identify the conception of time now pervading psychology and trace its influences. This is the purpose of the present book. Unfortunately, this purpose creates a problem because the influence of time spans virtually all the subdisciplines of the field. With rare exceptions (primarily introductory books), specialization in psychology has eliminated books covering more than one subdiscipline. Not only do publishers fear they will find no audience (i.e., specialists apparently only read books in their specialty), but potential authors also fear to venture outside their own specialties to write such books. I am, of course, one of those fearful authors. (I won't presume to speak for my brave publisher.) Still, I am confident that my lack of specialist credentials in *some* of psychology's subdisciplines does not invalidate the main analysis.

Like all authors, I must acknowledge my responsibility for this analysis. But like all authors, I must also acknowledge my dependence upon a host of others without whom this book could not have been written. I begin with my God and my wife, because they provided the inspiration and support needed to do the book. I hope that I honor them. I next turn to my colleagues and friends, whose support and guidance have been invaluable. I am especially grateful to three colleagues who have been helpful at every stage of the writing (and thinking): Joseph Rychlak, Richard Williams, and Brian Vandenberg. My "local" support has included my department chair, Helen Benedict, my kind associate, Mark Pantle, and the good graces of Baylor University. Finally, I wish to give a special note of thanks to Nick Bellegie, who first showed me that *non*psychologists could make sense out of this book, if not become greatly enthusiastic about it.

INTRODUCTION

When I was growing up, few political or even economic issues seemed to disturb the serenity of our small farming community. One issue, though, clearly sticks out in my mind. The farmers of our town were so concerned that they would have fought a war over it—if they could. Congress was again debating the merits of daylight–saving time. The Arab oil embargo was in full swing, and research had hinted at the possibility of widespread energy savings. All we had to do was turn our clocks up one hour. Nevertheless, the people of our community found this proposal positively abhorrent. I heard things like: "They have no right to fool around with God's time," and "They might *say* that it is 3 o'clock, but my cows and I will know that it is *really* 2 o'clock."

This was my first exposure to the concretizing of time. The farmers in my community had concretized *one* organization of events as if it were *the* organization of events. In point of fact, standard time was invented in 1883 to standardize railroad schedules. As long as the main mode of transportation was horse and carriage, towns and cities kept their own time. The speed of the railway system, however, required uniformity—hence, the birth of standard time. The people of my community had forgotten or never knew the human origins of this temporal organization of reality. Standard time had become so familiar, so "standard," that they had assumed it *was* reality—"God's time."

By the time I entered my chosen field of psychology, this incident had moved to the deeper recesses of my mind. I would never forget the tenacity with which my community had clung to its arbitrary temporal structure, but any consideration of it in retrospect seemed more curious than important. In graduate school, though, I realized that my community was not that unusual. Indeed, nearly everyone in psychology and everyone in our culture engaged in the same sort of concretizing of time. Daylight–saving time was not the issue, of course. A subtler and more pervasive organization of time was influencing our everyday thoughts about the world—*linear time*. The intent of this book is to begin a systematic discussion of this subtle organization in psychology.

Linear time is the metaphysical assumption that time flows like a line, independently of the events it supposedly contains. It advances uniformly forward as an invisible medium for all events. All events are distributed across this medium, and all scientific processes must occur along this "line." The three time dimensions—past, present, and future—are themselves thought to occur in a linear sequence, with the past and future separated by the present. As Arthur Lovejoy put it: "Time as ordinarily conceived is sundered into separate moments which are perpetually passing away. The past is forever dead and gone, the future is non-existent and uncertain, and the present seems, at most, a bare knife-edge of existence separating these two unrealities."[1]

This linear conception of time also places an interesting primacy upon the *past* (or the *first* in any sequence of events). Although the past is "dead and gone," to use Lovejoy's terms, it is also thought to have the most to say about the present and future. The linearity of the three time dimensions means that each must be consistent with the other. Because the past occurs *first* in the sequence, the present and future must remain consistent with it—as one uniform "flow." This is part of the reason that causality and time have been so closely allied. Being consistent with a "dead" past seems akin to being causally determined by that past. In either case, the past is our culture's primary means of explaining (and predicting) the present. Indeed, many people would question whether it could be explained any other way, and this may be the most telling evidence of linear time's influence.

Linear Time and Psychological Explanation

Most psychologists follow a similar approach to explanation. They hold, of course, that the mere chronological ordering of events does not ensure their causal relation. Nonetheless, this simple distinction is often violated. Sometimes just the occurrence of an event in a person's past is sufficient to endow that event with causal status in psychology. For example, the mere existence of a "childhood trauma" is enough for many psychotherapists to expect causal consequences in the person's distant future. Empirical tests may, of course, lead to the confirmation of such expectations. Still, many psychotherapists do not need a rigorous scientific test because the assumption of linear time leads so readily to this expectation. In fact, several scientific practices follow this expectation. For instance, the perfect correlation of a sequence of variables is often considered sufficient for causal inference.[2]

Linear time has made its greatest impact in psychology through its fusion with other concepts vital to explanation, such as causality, change, process, and behavior. All such concepts are thought to occur along the line of time. Indeed, many psychologists would be hard-pressed to conceive of such concepts *without* linear sequence, i.e., the "first" being the most influential.[3] The principle of causality, for example, epitomizes this linear sequence. Because the cause is thought to produce the effect, the cause is viewed as *necessarily preceding* the effect in temporal sequence. Change too is considered to be tracked through a sequence of observations along the line of time. The upshot is that linear time seems inextricably fused to explanation in psychology, and the *past* is the fundamental means of explaining psychological phenomena in the present. (The final section of Chapter 9 below shows how causation and change are conceived without the confounding of linear assumptions.)

Certainly, Freud is thought to have made psychology's dependence upon the past abundantly clear. Although his original ideas have been greatly modified, contemporary psychoanalysts have not changed this aspect of his theorizing. Childhood experiences are still viewed as crucial to present emotional problems. Even theorists that are diametrically opposed to Freud on many other issues seem to agree with him on the significance of the past. Cognitivists attend to past inputs and memory encodings; many humanists refer to long-lost opportunities for growth; and behaviorists consider reinforcement history and immediately preceding stimuli. *All* consider some form of the past to have primacy over abnormal, and even normal, behaviors in the present. To give the present or future this primacy (particularly over the past) would seem ludicrous. As we will see, conceptions that give this primacy to the present or future are often based upon nonlinear assumptions of time.

The primacy of the past is also carried into psychological treatment. The origin of abnormality is always the target of treatment, explicitly or implicitly. Psychological treatments attempt to undo, in one way or other, the effects of previous "doings" in time. All psychotherapists must work in the present, of necessity. But most focus therapeutic discussion on the patient's past for any number of reasons. Certainly, if the client's problems are caused by events in the past, then they can only be understood in light of that past. A patient's "history" is thus routinely taken. Moreover, psychological treatment must take into account this history. It is not uncommon for therapy to concentrate almost exclusively on either immediate or distant past events to obtain "insight" or garner some other form of cognitive or emotional understanding.

Interestingly, the widespread influence of linear time has *not* led to any widespread discussion of its influence. Issues such as psychology's dependence upon the past are rarely, if ever, examined or debated. Psychological researchers have investigated how people perceive the passage of time and how they better manage their time, but psychologists have not examined their *own assumptions* of time in their theories and explanations. Introductory books in psychology contain no references to time in this respect at all. Special-topic texts, such as books on cognitive, social, and clinical psychology, likewise indicate no allusions to temporal assumptions. Rare exceptions to this are virtually ignored in the mainstream literature.[4] A review of prominent journals and books reveals no research or ongoing discussion about the implications or role of time in the discipline.

Purpose of the Book

This begins to hint at the problems presented to psychology by linear time. While it may be true that this assumption has been very useful, it is also true that we have almost no knowledge of its influence on the discipline. The present book is intended to facilitate discussion of this influence. How is linear time currently affecting psychology's theories and therapies? Does it restrict or limit explanations in the discipline to some degree? Are there *non*linear theorists within psychology? If so, what views of time underlie their explanatory efforts? How might these alternate views help or hinder the task of psychology? Answers to these questions seem vital, especially given the pervasiveness of the assumption. In fact, it is difficult to imagine a topic of wider import for all aspects of the discipline.

The assumption is so widely held that an important clarification is needed: *It is not the purpose of the book to question the existence of time itself*. Linear time has become so confounded with time that the two are virtually indistinguishable for many people in Western culture. Our organization of time has been reified as "the way time is." This means that any challenge of the status of linear time is seen as a challenge of the status of time itself. This is not, however, the aim of the present book. Time *is* distinguishable from linear time—the former having to do with change, and the latter having to do with the organization or interpretation of that change. In this sense, the topic of the present book is the *organization or interpretation of change* by psychologists.

Part of the reason for the confounding of time and linear organization is time's so-called arrow. A growing number of scientists seem to be challenging the classical belief that time can run both "forward" and "backward"—i.e., time's *reversibility*.[5] They contend that change processes (time) run primarily forward. For example, we never experience broken windows reassembling themselves or people getting younger. We only experience the forward movements of these processes. Many scientists and lay persons have assumed this arrow of time indicates that time is linear.[6] This is unfortunate because even if time is like an arrow, the arrow does not have to follow a linear path. That is, the directionality of change is *not* prima facie evidence for linear time.[7] The past (or the first in the sequence) does not *have* to have primacy for such forward movement.

Consider as a contrast those theorists who assert the primacy of the *present* (or even the *future*[8]). Many holists and structuralists, for example, contend that present events have primacy when explaining present events.[9] Although the importance of the past and the directionality of events are not denied, these holistic theorists differ with linear theorists as to which time dimension exerts the most influence upon events. These holists look primarily to the events that are *simultaneous* (i.e., present) with the events under study rather than the events that *precede* them in linear time. Examples of important simultaneous events include: context,[10] system,[11] synchronicity,[12] interpersonal field,[13] and implicate order.[14]

A fundamentally different interpretation of time (change) is evident in these holistic explanations. This has led advocates of these explanations to challenge other characteristics of linear time as well, i.e., objectivity, continuity, universality, and reductivity (see Chapter 1). Similar to the primacy of the past (linearity), these other characteristics have been assumed to be characteristics of the change process itself. This assumption, though, confounds the linear organization of the interpreter with the directionality of the events themselves. The point is that directionality of events does not itself indicate which type of temporal interpretation is correct.

The question of how we understand and explain this directionality (and each of the time dimensions) still remains.

The present book addresses itself to this question. Unlike other recent books on time,[15] this work does not debate the irreversibility of time. The directionality of many (if not all) change processes is assumed, and the importance of all time dimensions—the past, present, and future—is unquestioned. Indeed, most chapters of the book include extensive discussions of the history of the issue at hand. What *is* questioned is the meaning and role of each of these temporal dimensions. Alternative interpretations of time are available that provide optional explanations and theories for psychologists. None of these alternatives dismisses the past or challenges our phenomenal experience of time's arrow. Nevertheless, few of these alternatives have even been *considered* by mainstream psychology, let alone evaluated for their worth to the discipline.

The problem is that the reification of linear time has led to a Kuhnian-type paradigm in psychology.[16] This means that linear time is an unconsciously agreed upon pretheoretical assumption. Psychologists have not only failed to discuss the merits and implications of this unconscious assumption, but they have also been unable to test it empirically because it is institutionalized in their method (see Chapter 4). How then do we facilitate greater awareness of this paradigm's impact, when it holds such axiomatic status within the discipline? Here, Kuhn's analysis of scientific paradigms is helpful. Paradigms typically first enter the awareness of scientists through *anomalies*—empirical findings and practical applications that "violate" the properties of the paradigm in question.

True to form, psychology manifests many anomalies to its own temporal paradigm. Many anomalous explanations are already recognized as having irregular or deviant properties. They are viewed as unscientific or mystical, when their primary difference from mainstream explanations is their implicit or explicit rejection of linear time. The difficulty is that each anomaly is embedded within its own subdiscipline of psychology. Anomalies are seen as isolated pockets of scattered resistance, rather than subtle variations upon a common theme. One of the purposes of this book is to cut across these specialties and pull together the major anomalies under one conceptual roof—namely, as anomalies to the linear time account. Consequently, anomalies are examined in the more basic contexts of psychology—development, personality, method, and cognition— as well as the more applied contexts—individual, group, and family psychotherapies.

Content of the Book

Before doing this examination, however, it is important that we understand what the characteristics of linear time are and how psychology specifically acquired this unrecognized assumption. Actually, linear time is a relatively recent conception. *Chapter 1* chronicles the cultural rise of this conception, and the reasons that psychology has not examined it critically. Crucial to this was psychology's modeling of Newtonian physics. Attempting to establish itself as a science, psychology modeled not only the scientific method of Newton's physics but also the philosophy of Newton's *meta*physics. Central to both was Newton's rendition of linear time— absolute time. Although later physical scientists were to challenge, if not reject, this assumption, psychology essentially retains it. All five characteristics of Newton's temporal framework for explanation—objectivity, linearity, continuity, universality, and reductivity— are shown to be alive and well in the explanations of psychology.

The chapters that follow (with the exception of the final chapter) attempt to describe how these characteristics have become the fundamental framework for explanation in each of several psychological specialties. Affected are not only the basic specialties, such as developmental and cognitive psychology, but also the more applied specialties, such as group and family therapy. Still, subtle anomalies to these characteristics have developed. The roots of these anomalies seem to stem from disciplines outside of psychology for the most part, but their influence is clearly evident in each specialty. Therefore, each of the chapters also reviews these anomalies to the linear paradigm. Their nonlinear properties are explicated, and their implications for alternative assumptions of time are explored (particularly in the final chapter).

Developmental psychology—the topic of *Chapter 2*—would seem the most dependent of all psychological specialties upon linear time. Developmental events are viewed as occurring *in* time, stages of development are thought to be connected *by* time, and rates of change are investigated as though an objective framework for their measure were self-evident. In fact, the historical rise of interest in developmental issues is linked to the cultural rise of linear time. What is perhaps surprising is the number of *non*linear developmentalists. These researchers have discovered anomalies to the linear paradigm, including dramatic "discontinuities" in development that deny the connections between early and later behavior, which a linear theorist would expect. Indeed, many of the so-called "continuities" of traditional developmental psychology are now considered to be "invented" through the linear bias of theorists or the linear methods of researchers.

Perhaps no field in psychology has more conceptual diversity than personality theory—the subject of *Chapter 3*. Although most contemporary personality theorists seem to desire linear characteristics in theorizing, classical theorists vary widely. Some theorists, such as Dollard and Miller, Skinner, and Bandura follow an objectivist approach to time. Time is considered to be an objective entity (in the tradition of Newton). Time's existence as an independent entity allows it to determine the objective arrangement of psychological events, and therefore the processes of personality and learning. Freud is often depicted as a temporal objectivist, but his writings also reveal considerable leanings toward temporal subjectivism. This latter approach denies time's objectivity. Time is viewed as one of the ways in which the mind imposes its own order (subjectively) upon events of the world. Jung and Lewin are shown to be anomalous theorists in this subjectivist tradition. They place more emphasis upon the present or future for explanation. The past is by no means ignored, but it is not viewed as having linear primacy or being independent of consciousness.

A major reason for the strength of linear time in psychology is its perceived necessity for method—the topic of *Chapter 4*. When linear time is considered necessary to method, any idea tested by that method is likely to be construed in a linear fashion. Similar to Newton, psychology's best research strategies—its experimental designs—depend upon the linearity of the phenomena being studied. Moreover, justifications for the use of laboratory and simulation methods, as well as current notions of replication, all depend upon characteristics of linear time. Interestingly, historic approaches to method did not require these characteristics. Two contemporary anomalies to linear methods appear to draw upon these historic approaches to some degree. Systemic methods originate from the natural sciences, whereas hermeneutic methods originate from the humanities. Both approaches reject most, if not all, of the linear characteristics of method considered so essential to current mainstream methods.

Cognitive psychology—the topic of *Chapter 5*—is considered to have revolutionized psychology from its obviously linear, stimulus-response roots. Has this "revolution" altered the influence of linear time? Not only does the answer appear to be no, but it also seems clear that the cognitive revolution is instrumental in *retain-*

ing Newtonian assumptions of time. Cognitive explanations have essentially updated the mechanistic metaphors of Newton based on the clock to those based on the computer. Although more emphasis is placed on unobserved "software," our computer-like minds are thought to be the modern preservers of the past and its linear primacy. Recent investigations by rationalistic "constructivists," however, indicate that we *re*interpret or reconstruct our memory in light of what our mental set is in the *present*. In this sense, it is more accurate to say that the present causes the meaning of the past, than it is to say that the past causes the meaning of the present. Piaget's interactionism, Rychlak's learning theory, and Dreyfus's critique of artificial intelligence are discussed in this light.

A three-chapter survey of the various modalities of psychotherapy—individual, group, and family—begins with individual psychotherapy in *Chapter 6*. The prominence of individual therapy is shown to be related to the reductive characteristics of linear time. As linear time gained prominence historically, people were viewed more in relation to *themselves as individuals* in the linear past, than in relation to *each other* in the synchronous present. This chapter also describes the applications of many of the objectivist and subjectivist personality theories of Chapter 3. The vast majority of therapeutic strategies—at least as formally espoused—operate under objectivist (and linear) assumptions of human nature. Nonetheless, a small minority of anomalous therapists explicitly conceptualize therapy "outside of time." This means that the emphasis is upon logical rather than chronological significance, and "first" does not imply "fundamental," as an objectivist would contend. If anything, fundamental issues in therapy stem more from relevance to the present and expectations of the future.

As *Chapter 7* on group psychotherapy demonstrates, the presence of more people in therapy does not necessarily mean a different conception of therapy. Therapeutic individualism remains quite strong in conceptualizations of group therapy. Considerable emphasis is often placed on linear and reductive therapeutic techniques. Because a person's unique past is considered to be primary to a therapist's understanding and treatment, the individual (rather than the group) is the therapist's focal point. Indeed, many group therapies are more accurately viewed as individual therapies with the rest of the group as an audience. Anomalous group treatments stem primarily from so-called here-and-now approaches. Sometimes the phrase "here-and-now" merely signals an emphasis upon the immediate past (e.g., recent problems), but some here-and-now strategies truly endow the simultaneous present with primacy. Because the bonding and experiencing of the group members occur in the now, this approach to group therapy declares that the therapeutic focus must also be in the now.

Family therapy—the topic of *Chapter 8*—is a field of psychology that has dedicated much of its theoretical resources to the formulation of an alternative to linear causation. Still, many of these "alternatives" have overlooked the source of causality's linearity—linear time. In communication approaches to family therapy, for example, the doctrine of causality has been modified to "circular causality." Unfortunately, as the chapter describes, this modification has not avoided causality's reductionistic qualities. Because the system is thought to occur piece-by-piece along the line of time, the system-as-a-whole is never present to the therapist at any point in time. Family therapists are thus resigned to interventions that *directly* affect only a reduced portion of the system. Anomalous family strategies, on the other hand, conceptualize interventions that affect all parts simultaneously because significant aspects of the system exist concurrently. Structuralists, for example, value present-focused interventions into nonreducible family units—without piece-by-piece sequence.

Chapter 9 seeks to distill the essence of the Newtonian legacy for psychology from the previous chapters. If linear time is an unacknowledged paradigm of psychology, what are its implications for explanation in general? What does it rule in and out of the discipline? This chapter attempts to answer these questions by explicitly listing all the essential manifestations of the Newtonian paradigm for explanation. Each manifestation is described and related directly to the various subdisciplines of psychology. This is followed by a similar summary of *anomalous* explanations and their relations to specific psychological specialties. Do these anomalies require the abandonment of causation and change? Despite the historic grafting of linear time onto these constructs, the remainder of this chapter shows how causation and change are compatible with most anomalies, once this graft is removed. A broader forum for the discussion of explanation in psychology is thereby provided.

The purpose of *Chapter 10*, the final chapter, is twofold: sketch alternate assumptions about time, and provide some indication of their implications for psychology. This brief sketch of alternatives will not be sufficient to begin new paradigms or competing schools of psychology, and this is not its intent. Alternatives are provided primarily as a contrast to linear time. The axiomatic status of this assumption makes it difficult to identify and evaluate without other assumptions from which to distinguish it. Therefore, with the anomalies of the previous chapters as our guide, two temporal assumptions—*organismic holism* and *hermeneutic temporality*—are outlined. The second task of this final chapter is to bring these less familiar assumptions to life in psychology. Consequently, implications for each of the subdisciplines of the previous chapters are briefly outlined.

Chapter 1

NEWTONIAN TIME AND PSYCHOLOGICAL EXPLANATION

Linear time pervades psychological explanation. As outlined in the Introduction, linear assumptions are foundational to both the basic and applied aspects of psychology. How intriguing it is to note that this issue has virtually never been discussed. With rare exceptions, time has not only managed to avoid systematic examination, it has scarcely been acknowledged anywhere in the field. What is the reason for this lack of acknowledgment? Psychology is a scholarly discipline with many highly educated people examining its ideas and explanations on a continual basis. How could a metaphysical assumption of this importance escape some sort of scrutiny?

As this chapter shows, two factors appear to be the most responsible for this lack of scrutiny: psychology's past and psychology's future. Regarding its history, psychology is a relatively new discipline. Although many ancient thinkers have certainly influenced psychology, psychology's identity and style of explanation are distinctly modern (and "modernist") in outlook.[1] Many metaphysical assumptions are implicit to our modern culture, and contemporary psychologists have adopted some of these assumptions without full awareness. As this chapter describes, one of these cultural assumptions is linear time itself.

Regarding psychology's future, there is no mistaking the early goal of psychologists—being "scientific." Wilhelm Wundt, for example, titled perhaps the first psychology course in 1862, *Psychology as a Natural Science*. The only question was: How does a new discipline become a natural science? Here, psychologists seemed to adopt the very reasonable approach of looking to other successful sciences as their guide. This chapter outlines how Newtonian physics—the ideal of sciences during psychology's formative years—became a prime model for all that early psychologists wished to become. Much of this modeling, however, involved Newtonian assumptions of the world that were accepted uncritically. One of these assumptions, as this chapter reveals, is Newton's own rendition of linear time—absolute time.

The bulk of this chapter outlines how Newton's conception of time has influenced scientific explanations ever since. As we shall see, five characteristics of explanation have devolved from Newton's temporal framework: objectivity, continuity, linearity, universality, and reductivity. The chapter first describes each of these characteristics as it is related to absolute time. Then, criticisms of this framework by philosophers and physicists are discussed. The latter group of critics is especially important because many contemporary physicists have abandoned these temporal characteristics (as well as absolute time) in their explanations. Nonetheless, contemporary psychology—with its identity now somewhat intact—has not looked back to physics. This chapter indicates how psychology currently maintains most, if not all, of these linear characteristics in its mainstream explanations of behavior, mind, and abnormality.

The Rise of Linear Time in Western Culture

As noted, the view of time held by so many psychologists is also the view of time held widely in Western society. In fact, time is not typically seen by lay culture as a "view" at all. It is "out there," flowing like a line from past to present to future. No examination of this linear notion of time is considered necessary because it is part of reality. Past events are viewed as fundamental to explanation, just as in psychology. Indeed, it is considered common sense in Western culture that our personalities and attitudes are caused by our past experiences. The question is: How did this common sense become so common? How has this linear view of time gained such a hold on our culture and such an authority in psychology?

Actually, the predominance of linear time is a relatively recent phenomenon. Ancient peoples did not view time as an objective frame of reference for marking events. They relativized time by making it conform *to* events, rather than events conform to time. For the Romans, each hour of daylight in the summer was longer than each hour of daylight in the winter. Time was a dynamic and adjustable organization tailored to fit our world experiences. Cyclical, rather than linear, views of time dominated these cultures because so many aspects of nature seemed cyclical, such as the seasons and heavenly bodies. Plato believed that the order of world events was destined to repeat itself at fixed intervals. Aristotle's students wondered whether Paris would once more carry off Helen and thus again spark the Trojan War.[2]

Our Western view of time arose primarily as a result of three historical developments: the spread of Christianity, the industrialization of society, and the invention of cheap watches.[3] Pre-Judaic religions complemented the cyclical view of time. They portrayed time either as infinite and possessing no beginning or end, or as a cycle of rebirth and future life with time forever repeating itself. The spread of Christianity brought to bear a "stunning" new conception.[4] Christians considered their God to be the creator and ultimate destroyer of the universe. Hence, the world had a beginning and an end, and important Christian events, such as the birth of Christ, were unique and nonrepeatable. The spread of these conceptions resulted in a competition between the cyclical and linear views during the medieval period.[5,6]

The temporal tide began to turn in the favor of linearity—at least for our Western culture—when industrial economies arose. As Lewis Mumford concludes, "The clock, not the steam engine, is the key machine of the modern industrial age."[7] When power stemmed from the ownership of land, time was considered plentiful and cyclical, being associated with the unchanging cycle of the soil. With the rise of a mercantile economy and the mechanism of industry, however, emphasis was placed on the scarcity of time and "forward" progress.[8] The byword became "Time is money," and implied that time could be saved or spent.

The coup d'état for the linear view was the increased availability of cheap watches. The mass production of watches in the nineteenth century made it possible for linear time to regulate even the most basic functions of living. "One ate, not upon feeling hungry, but when prompted by the clock: one slept, not when one was tired, but when the clock sanctioned it."[9] Regulation of our lives by the clock meant that the abstract assumption of linear time could be endowed with a sort of concrete reality.[10] People now seemed to be able to "see" and "feel" time (the clock). Time also appeared to be one of the causes of psychological factors because the thoughts and behaviors of individuals seemed to turn on what time "told" them. In short, a convenient (linear) way of organizing events became reified as *the* way events were organized.

Psychology was conceived and developed during this temporal zeitgeist; time was a concrete actuality, rather than a point of view. The spread of Christianity, industrialization, and the invention of cheap clocks, all coalesced to make linear time a "reality." Before this coalescence, many scientists, such as Newton, felt it necessary to make their assumption of time explicit. Several views of time were possible,[11] and so one's view had to be identified and supported. Psychologists, on the other hand, were not called upon to articulate their temporal assumptions. Linear time had become a given and required no discussion or defense. Time existed like a line, independently of us, and virtually everyone accepted this reification without awareness.

Newtonian Time

Psychology is not the only discipline to have reified time. Einstein found a similar state of affairs in his own discipline of physics. The quotation from Einstein that serves as this book's epigraph evidences this: "Concepts which have proved useful for ordering things easily assume so great an authority over us that we forget their terrestrial origins and accept them as unalterable facts."[12] Einstein's point here, of course, is that sometimes the very pervasiveness of an idea leads to its anonymity. Certain ideas can be so commonplace and so widely accepted that they go completely unrecognized. Yet it is these very ideas that are often the most influential to thinkers in a discipline.

One of the ideas that Einstein was alluding to in this quotation is the idea of time itself. Part of his genius was the realization that time played an unrecognized role in physics. Indeed, linear time was seen as an absolute truth—an unquestioned part of reality—during the preceding three hundred or so years of physics. There was no reason to examine its "role" because it was not even viewed *as* a view. This led to a curtailment in the number of new ideas in physics.[13] Acceptable ideas about reality had to be compatible with time's supposedly linear properties. Einstein's theory of relativity, however, was largely based upon his examination and eventual rejection of this traditional view of time.[14] He proposed an alternative view that ultimately revolutionized the discipline of physics in the twentieth century (discussed later in this chapter).

Still, this revolutionary view has had little impact upon lay culture. Except for parts of physics and philosophy, the Newtonian picture of the world is by far the dominant one in Western culture.[15] Newton's views are now the common sense of our culture in regard to time. Many historians give Newton an authority "paralleled only by Aristotle" in his influence on Western society.[16] Although his basic ideas probably originated with his intellectual ancestors,[17] Newton gets the credit for assembling the ideas into the current package our culture calls time. The distinguished philosopher and historian, Edwin Burtt, put it this way:

> Magnificent, irrefutable achievements gave Newton authority over the modern world, which, feeling itself to have become free from metaphysics [such as time] through Newton the positivist, has become shackled and controlled by a very definite metaphysics through Newton the metaphysician.[18]

Newton's views have had a similar effect upon psychology. A number of historians and other scholars have noted that psychologists considered Newtonian physics their model of science during psychology's formative years.[19] At the time of psychology's inception, Newtonian physics was the queen of sciences. Modeling Newtonian physics was only natural for a new discipline struggling to become a science. Psychologists took not only their principles of explanation from Newton but also their approach to scientific method. Newton's views of time—views that were later to be challenged, if not rejected, by many later physicists—were implicit to everything adopted by fledgling psychology. And because these views were never subsequently examined, psychologists continue to employ them in close to their original form. Therefore, let us examine Newton's views in more detail.

Newton postulated *absolute time* which "of itself, and from its own nature, flows equably without relation to anything external."[20] Newton needed this assumption for two main reasons. First, his conceptions of motion and causality required an absolute frame of reference.[21] Motion, for example, could not be detected or measured without an objective past and present. The rolling ball begins its roll at some point in the past, but is now at some point in the present. Second, his mathematics required the continuity of events (flowing equably). He regarded moments of absolute time as a continuous sequence like that of real numbers, believing that the rate of this sequence was independent of the events taking place in them.[22]

For these reasons, absolute time became the standard by which all scientific explanations were judged. The order (and directionality) of the world was thought to be synonymous with the absolute and linear organization of events. Characteristics of Newton's absolute time became the "rules" for acceptable scientific explanation for nearly three centuries, and still form the rules for many disciplines such as psychology. It is thus important that we explicate these rules and their modern criticisms, and then check the specific role these rules play in psychological explanation.

Newton's Temporal Framework for Explanation

Newton's approach to time left science with a legacy of five somewhat overlapping implications or characteristics for scientific explanation. These include objectivity, continuity, linearity, universality, and reductionism. Some of these characteristics are the properties of time itself, as envisioned by Newton, and some are the necessary properties of the events to be explained, because they are *in* absolute time.

The assertion that events are "in" time is itself an implication of a temporal characteristic. Newton viewed time as *objective,* existing "absolutely" and independently of consciousness. Time is conceived as a medium *in* which and *against* which events occur and can be related to one another. Motion, causation, and change are seen to exist "out there," and so an absolute framework for evaluating these conceptions must also exist out there, separate from them (and our consciousness). If time were subjective—Newton might argue—distinctions between the temporal dimensions (past, present, and future) would be left up to the perceiver, and an objective science would be in jeopardy. Indeed, the notion that cause and effect require succession in time occurred with the advent of absolute time.[23]

This view of causality was bolstered by another property of Newtonian time, its *linearity*. Just as a line is thought to consist of a succession of points in space, time is considered to consist of a succession of moments in time.[24] The three dimensions of time—past, present, and future—thus occur in a linear sequence. Time begins in the past and advances into the present on its way to the future. (Absolute time is *not* "reversible."[25]) This places the greatest weight upon the past (or the "first" in a sequence), because it is the temporal entity that supposedly starts this process. The metaphor of the line means that the present and future must remain consonant with the past. Moreover, the past is the temporal entity with the most utility. The present is less useful because it is just an evanescent "point" on the line of time, and the future is less useful because it is not (yet) known with any certainty. Only information from the past is viewed as substantive and certain enough to be truly known and understood.[26]

Newton also considered time to be *continuous,* proceeding smoothly and "equably," as he put it.[27] Actually, this characteristic of time has two properties that are worth separating: consistency and uniformity. Consistency is the well-known Newtonian notion that events which happen at one point in time are consistent with events occurring later in time—the past is continuous with the future. This is the origin of Newton's conviction that the world is predictable. If enough is known about the present situation (or the past), then future events or states can be predicted. Uniformity, on the other hand, is the notion that time is homogeneous. Although the events *in* time can move at different rates, time does not itself slow down at some points and speed up at others—it "flows" at a constant, never-changing pace. This uniformity provides the perfect frame of reference for measuring events.

Time's continuity has significant implications for change. In Newton's metaphysic, change can not be discontinuous or instantaneous, moving abruptly from one state into the next.[28] Change has to be continuous and smooth, much as a flower gradually blooms. The reason is that Newton conceived of time as *infinitely divisible*—like a line. No matter how small the interval of time, there is always a line of time (points in time) that spans the interval. This means that change can only be incremental. Whatever change occurs, it is assumed to have intervening levels that correspond to the intervening points in time. Change can occur at different rates, and motions can proceed faster or slower. However, change can not occur through sudden jumps from one stage into another—such as a flower bud jumping to a full bloom—without some points of time (and levels of change) falling *in between* the two stages.

This characteristic of continuity has led to another major feature of scientific explanation, labeled by some authors as "universality,"[29] "atemporality,"[30] or "symmetry."[31] This characteristic of *universality,* as we shall call it here, assumes that natural laws are universal and unchangeable, regardless of the period of time in which they are observed. Natural processes are still thought to unfold across time in the continuous manner just described. Nevertheless, the principles behind the processes are considered to be independent of the events and particular period of history in which they unfold. The laws of planetary motion, for example, are the same laws at one point in earth's history as they are at another point in earth's history. This universality is only possible if time is uniform in the Newtonian sense. If time changes its rate or quality, then the temporal relations between planetary events would not be consistent from one period of history to the next. Scientific laws, in this sense, would not be lawful.

The notion that lawful processes take place across time has had another implication for explanation—*reductionism*. Reductionism results from the fact that any one moment in time contains only a reduced portion of the process. That is, if a process begins at Time 1, continues through Time 2, and ultimately culminates at Time 3, the process *as a whole* literally never exists. Only part of the process can occur *at any one moment* in time. Recording devices, such as an observer's memory, permit a part of the process to be "photographed" and juxtaposed with the next moment's part until all the process is viewed *at the same time*. However, no *direct* access to the process-as-a-whole is ever possible. (A memory of previous parts is not direct access.)

This also makes interpretation of each part's relation to the whole problematic because each part crosses our window of the present independently of the whole. Any properties of the part that may be derived from its relationship to the process-as-a-whole are not available. Without these properties, an understanding of the process-as-a-whole is itself problematic. All that is available at the end of the process is the cumulative record of independent parts, as each part is encountered in time, and not information about how these parts are related as a whole.

Newton brilliantly coalesced all five characteristics of explanation into a coherent package by calling upon *mechanistic metaphors*. He felt the universe—with its motions and chains of causation across time—was directly analogous to the great machine of his day: the clock. Through his writings and research, he combined the implications of absolute time just described—objectivity, linearity, continuity, universality, and reductionism. He represented them all with machine metaphors that seemed to embody these characteristics.[32] Machines seem to operate objectively through a continuous and linear sequence of events. This sequence is universal because it appears to be repeatable, regardless of the period of time in which the repetition occurs. Machines also seem to evidence temporal reductionism in their functioning. Their sequentiality provides no direct access to the whole of their processes at any given moment in time.

When the universe is presumed to possess these five temporal characteristics, explanations that are properly "scientific" also possess these characteristics. Mechanistic explanations of data are, of course, preferred because they naturally embody these characteristics. The reverse is also true—those processes that manifest linear and lawful properties are considered mechanisms and thus accorded appropriate scientific status. Newton even carried his temporal approach to explanation into his *method*. In order to observe parts of the machine universe in its mechanistic regularity, he assumed that one tracked the effect of some antecedent (in time) experimental manipulation on its consequent. Orderly relationships between variables can thus be observed and cataloged until all the universe is understood.

Criticisms of Newton's Framework

As undeniably brilliant and influential as this temporal framework for explanation has been, it has not avoided criticism. G. J. Whitrow, for example, characterizes Newton's conception of time as the "most criticized, and justly so, of all Newton's statements."[33] Many subsequent philosophers and physicists have called Newton's conception into question on theoretical, practical, and empirical grounds. For example, Whitrow notes that the "equable flow" of time is problematic on purely theoretical grounds:

> If time were something that flowed then it would itself consist of a series of events in time and this would be meaningless. Moreover, it is equally difficult to accept the statement that time flows "equably" or uniformly, for this would seem to imply that there is something which controls the rate of flow of time so that it always goes at the same speed. However, if time can be considered in isolation "without relation to anything external," what meaning can be attached to saying that its rate of flow is not uniform? If no meaning can be attached even to the possibility of non-uniform flow, then what significance can be attached to specifically stipulating that the flow is "equable?"[34]

Some have questioned the practical utility of Newton's conception of time as a frame of reference.[35] Because Newton regarded time as uniform and infinite, any position that an object might take *in* time is not discernible from any other position. One portion of time is identical (and uniform) to another. Wherever the object resides (in time), there is no distinguishing feature for that period of time, and there is a similar quantity of time surrounding it in the past and future (infinity). It is therefore impossible to locate an object in absolute time and establish whether it is in motion. Temporal position and motion can only be discerned with reference to another body (e.g., a clock), and Newton's conception of absolute time is unnecessary. Indeed, Newton's conception seems useless for the main reason he formulated it—as a standard for temporal position and motion.

Other criticisms of absolute time are longstanding, and convince most analysts that Newton was "mistaken in several different respects,"[36] or "uncritical, sketchy, inconsistent, even second-rate" as a theoretician.[37] The ancient philosopher Zeno, for instance, provided a penetrating critique of the infinite divisibility and continuity of time.[38] Other critics have focused upon Newton's confounding of linear flow (his theory) and temporal sequence (his data).[39] As noted in the Introduction above, the existence of temporal sequence—"time's arrow"—does not necessarily imply the existence of linear flow—the "first" being the most important in this sequence. Newton, though, considered all physical events to be influenced by the temporal medium in which they supposedly occurred. Therefore, any sequence of related events supposedly involved all the characteristics of absolute time described above.

The trouble is that a sequence of physical events does not *have* to involve these characteristics. Consider the sequence of hydrogen and oxygen gases becoming water. Although this particular set of events has a very definite and predictable relationship, this relationship does not have to be viewed as linear. That is, its predictability is not derived in classical Newtonian fashion from its "past." The past properties of hydrogen and oxygen gases do not permit us to predict the qualitatively different, future properties of water.[40] The predictability of this relationship stems from our repeated observations of this sequence, *not* from its continuous unfolding from a past state. In fact, this particular change (gases into liquid) can be construed as *dis*continuous in nature—from one qualitatively different gestalt to another. The point is that the directionality or sequence of natural events does not require linear or continuous characteristics (or any of the other characteristics of Newton's framework).

Newton, however, extended this confounding of linear theory to his method. Some philosophers, for example, have criticized him for "making a metaphysics of his method."[41] That is, Newton confused his metaphysical theory of the universe (being a linear and continuous machine) with his scientific method (observing the natural order of variables). He experimentally intervened in antecedent events to observe their later effects in time, all the while assuming that linear flow was involved in this sequential relation. In this way, his metaphysics could not be proved wrong. His method (sequential observation) made it seem that his assumptions of time were constantly being affirmed. If, on the other hand, a crucial event for explaining a phenomenon were *simultaneous*, Newton's linear method would be unable to discover it. Such nonlinear explanations would be overlooked owing to the institutionalization of linear explanation in his scientific method (see Chapter 4).

The most significant criticism of Newton's notion of time has come from his fellow physicists. Einstein's conceptual forerunner, the physicist Ernst Mach, criticized the reductive implications of Newton's conception, focusing particularly upon what absolute time did to causality. Mach felt that a linear conception was incapable of embracing the multiplicity of relations in nature. He viewed events of the world as *functionally interdependent*, with no particular event taking precedence over the other just because it occurred before the other in time. He noted that measures of time were themselves based on space, such as the spatial positions of clock hands or heavenly bodies. "We are thus ultimately left with a mutual dependence of positions on one another."[42] In this sense, our dimensions of reality are not time and space but space and space. There is no separate temporal entity against which to measure the past or future of even causal events.[43]

Einstein was also highly critical of Newton's temporal framework. In what follows, Richard Morris summarizes the effects of relativity theory upon absolute time:

> Time is not absolute, it is relative. As the special theory of relativity shows, time measurements depend upon the state of motion of the observer. Time is not a substance that "flows equably without relation to anything external" [Newton's assertion]. According to the general theory of relativity, the presence of matter creates gravitational fields that cause time

dilation. Finally, if time does "flow," . . . the movement of the "now". . . seems to be a subjective phenomenon. . . . At best, one can only say that time moves onward at the rate of one second per second, which is about as meaningful as defining the word "cat" by saying "a cat is a cat."[44]

Central to Newton's view is the notion that events which are simultaneous for one observer are simultaneous for all observers, regardless of their frame of reference. In other words, a particular instant of time is the same instant of time everywhere in the universe and, hence, absolute or universal. Einstein, however, demonstrated that this is not true through his special theory of relativity. Avoiding Newton's linear methodology, he used *gedanken* ("thought") experiments to show that two or more observers in relative motion do not necessarily agree that two independent events are simultaneous. When events A and B are simultaneous in one inertial frame of reference, A can be observed to occur before B in another inertial frame of reference. Moreover, B can be observed to occur before A in still another inertial frame of reference!

If one assumes an absolute temporal frame of reference, the next question is which observer is *really* correct? This query implies that only one (objective) interpretation of events is correct because there is supposedly only one temporal measure of events. The same events cannot occur in opposite sequences when observed at the same time. Nonetheless, Einstein held that *all* observers are correct within their own inertial frames of reference, and no observer is more correct than any other.[45] In short, there is no absolute truth about the matter. Einstein resolved the apparent contradiction between these observations by noting that time flow is not totally a result of the events themselves. The apparent flow of time is due, at least in part, to each observer's inertial frame of reference.[46]

Modern physicists have not only disputed the reductivity, linearity, and objectivity of time, they have also challenged the continuity of events across time. Many quantum physicists, for instance, contend that electrons move from one orbit to another instantaneously (without time elapse).[47] Electrons simply disappear from one quadrant and reappear in another. Similarly, changes between various stationary states are considered to be discrete and discontinuous.[48] Discontinuous change, as mentioned above, is akin to a flower growing from a bud to a full bloom instantaneously—one instant it is closed, the next instant it is fully opened. This seems to fly in the face of our linear notions of common sense. Our usual notion of time implies that one instant has to be connected to the next with a line, and thus there must be a small interval of time in which the change occurs. Nevertheless, quantum physicists have demonstrated that change can truly be discontinuous—not just faster rates of change but change without temporal duration.

Psychology's Newtonian Framework

These challenges to Newton's temporal framework for explanation have not been widely recognized. Linear time continues to reign supreme in our lay culture and most disciplines other than physics and philosophy. Linear time certainly rules unopposed in mainstream psychology. Psychologists modeled physics just before Einstein's revolution at the turn of the century, and never looked back. Because of the cultural factors described above, early psychologists never concerned themselves with conceptions of time. Linear time was assumed to be part of reality. Criticisms of Newton's temporal assumptions prompted no reexamination in psychology, because no temporal assumptions were even recognized. Psychology's reliance upon Newtonian assumptions, therefore, remains undeterred in virtually every important respect.

First, psychologists view time as happening *objectively,* existing independently of human consciousness. As Ornstein notes, "Most psychologists, in considering time, have taken for granted that a 'real' time, external to our construction of it, does exist, and that this time is linear."[49] Faulconer and Williams also discuss psychology's "objectification" of time,[50] and McGrath and Kelly observe that most research on time is "done on the premise that there is a singular, and known or knowable, objective time."[51] Many psychological experiments, for example, have been conducted to discover how accurately such "real" time is perceived. Time is treated as if it consists of its own stimuli for perception, though real time is always identified with clock-time. The clock, of course, only marks or measures time; the clock is not time itself. To call the clock "real time," as Ornstein points out, "is somewhat like calling American money 'real money:' it is parochial at best."[52]

Second, time is viewed as *continuous.* Psychological events are seen as continuous in the sense of later events being consistent with earlier events. Abrupt "discontinuous" shifts that are incongruent with previous events are thought to be improbable, if not impossible. As the developmentalists, Emde and Harmon have observed, most researchers have "expectations for connectivity and continuity,"[53] presuming a "linkage from early behavior to later behavior."[54] People in general are presumed to be continuous with their upbringing. Personalities and attitudes are traditionally thought to be consistent with the person's past experiences. Any behaviors or thoughts that appear to be exceptions to this rule (sometimes deemed "abnormal") merely indicate that some of the person's past is not known. If it were known, then we would see its continuity to the "exceptional" behaviors and thoughts in question.

Temporal continuity is also used to explain change in psychology. Indeed, in accordance with Newton, change and time are virtually synonymous—both being smooth and gradual. Change from one psychological stage to another must occur through intermediary states (or moments in time). "Spurts" of change are possible, but *some* amount of time must occur *between* changes. For example, changes that researchers consider "discontinuous" are often observed in child development.[55] Still, these are normally viewed as rapid continuous changes—changes across a short span of time— rather than changes with no time or transition between events.[56] A child cannot move from one stage of development to the next without passing "in between." Continuity implies that one instant is connected to the next with a line, and thus there must always be a small interval of time in which change occurs.

Virtually all psychological explanations are *universal*.[57] Psychologists have long sought general "laws" of behavior that are independent of the particular historic situation in which they are embedded.[58] Examples are Fechner's law of the strength of sensation and Skinner's principles of reinforcement—both presumably applicable today, despite their having been formulated many years ago. Most psychologists attempt to look "behind" their data to find the universal principles that underlie them.[59] Cognitive psychologists, for instance, study memory as if principles can be gleaned from experiments that apply uniformly to the memory of all persons in the specified experimental conditions.[60] These psychologists implicitly assume that time itself remains uniform from situation to situation.

The *linearity* of explanation in psychology is also readily apparent. Time is considered to "flow" across psychological events like a line, and distribute psychological processes into linear sequences.[61] This is most clearly observed in the "causal" explanations of psychologists. Any event observed "before" is automatically considered for, if not awarded, causal status over events observed "after."[62] Time intervals or points on the time line between cause and effect must be filled with causal process.[63] From this perspective, it is easy to see why so many psychologists place so much emphasis upon the past. The present is an effect of the past. Moreover, the present is only one point on the line of time, and a durationless and fleeting point at that. A person's life, therefore, consists of the past almost exclusively. It seems only logical that the most theoretical and therapeutic attention is paid to the past.

The fact that psychological processes supposedly take place across time has the same implication it had in Newtonian physics: *reductionism*. No process—whether it be mental, emotional, or behavioral—can exist as a whole at any point in time. A reduction of the process is all that is ever *directly* available for study. Consequently, it is only natural to conceptualize processes as component parts that are separated by linear time. Consider, for example, some models of family therapy. Although family therapists typically wish to conceptualize the whole of the family system, their theorizing often depicts this system as occurring piecemeal along the line of time (see Chapter 8). This type of linear explanation has overlooked reductive ramifications. Because the system as a whole is never present to the therapist at any one point in time, the therapist is resigned to interventions that directly affect only a portion of the system. No truly systemic intervention—at least in the sense of affecting all parts simultaneously—is possible from this Newtonian perspective.

Psychologists also seem to favor mechanistic *metaphors* for explaining psychological processes. As Anthony Aveni rightly declares, "Machinery is, for us, the power tool of metaphor."[64] Just as Newtonian explanations relied upon the clock, psychological explanations have historically relied upon a host of different machines. The human mind, for instance, has been analogized to whatever mechanism was ascendant in that day, from the hydraulics of the steam engine to the relays of telephone switchboards.[65] Today, of course, the computer is the ascendant machine, and true to form, computer metaphors abound in theories of the mind. Even families are understood through computer metaphors.[66] Computers, no less than their mechanistic predecessors, operate across time in temporal stages that minimally included input and output.[67] In this sense, Newtonian time and mechanistic models have served to catalyze the popularity of the other.

Finally, contemporary psychology and Newtonian physics conceptualize scientific *method* with similar temporal assumptions.[68] Psychological scientists view themselves as intervening experimentally, and then observing the consequences of this intervention later in time. This is aided by psychology's decidedly linear approach to causation. Temporal sequence is so conflated with causation that the two are often indistinguishable in research. Psychological experimenters have rarely been accused of "making a metaphysics of their method"[69] (as has Newton), but this may be due to psychologists not making explicit their own assumption of time. Without an awareness that linearity is a part of psychology's metaphysic, psychological researchers cannot be accused of confounding this assumption with their method. Yet, their method may incorporate linear time in a way that prohibits any true test of its validity.

Conclusion

It is important, then, that we identify the linear view of time in all its manifestations. Temporal assumptions cannot be discerned with a method that assumes them. Thus, the process of identification has begun in this chapter with a brief cultural and historical analysis. Our cultural analysis finds psychology's metaphysic to be a product of modern Western culture to some degree—likening time to a continuous line that is independent of the events it supposedly measures. Still, it is unlikely that psychology would have adopted this belief without reputable scientists also endorsing it—hence, the significance of Isaac Newton. Newton, to his credit, made his conception of time explicit. However, there is no indication that early psychologists (particularly those pressing for natural science methods) did likewise. Instead, methods and modes of explanation were adopted that implicitly contained Newtonian temporal assumptions.

Five of those implicit assumptions are delineated. Time is assumed to be *objective*, independent of psychological processes. Time is considered to be *linear*, with the past as primal and the other two time dimensions as following in consistent sequence. Time is also *continuous*. Because all events occur along the medium of a line-like entity, the world takes place in a uniform and smooth manner, ruling out precipitous jumps or cataclysmic changes. Such continuity allows for ultimate predictability and *universality*. Any lawful empirical process can be counted on to retain its original temporal relationship because the passage of time does not alter its quality. The only potential drawback is a lack of direct access to the process as a whole. Only *reduced* pieces are observable as they present themselves across our only portal to the world—the present.

It should not be surprising to find that all five characteristics of Newton's temporal framework are endemic to psychological explanation. Indeed, these five characteristics have served historically as an important guide to scientific explanation in general. Nonetheless, it would be inaccurate to assume that these characteristics have permeated *all* aspects of explanation. As described, modern physics has all but abandoned many of these characteristics. In psychology too, many intriguing anomalies to the linear paradigm have arisen in various subfields. The problem is that some psychological explanations have been considered to be linear when they were actually not, whereas other explanations have been thought to be nonlinear when they were really linear. It is therefore important to examine these "anomalies" to the linear paradigm of psychology. This is accomplished by reviewing the major temporal assertions of several psychological subfields in the next seven chapters.

Chapter 2

DEVELOPMENTAL PSYCHOLOGY

On the face of it, developmental psychology would seem the most dependent of all psychological specialties upon absolute time. After all, the main task of developmental psychologists is to study the person *across time*.[1] Developmental events are viewed as occurring *in* time, stages of development are thought to be linked *by* time, and rates of change are investigated as though an objective framework for their measure were self-evident. Indeed, many developmentalists would argue that an objective standard for time is utterly crucial to the study of development.[2] The very notion of development, as espoused by most theorists, seems to imply a person's psychological growth and physical maturation across a continual flow of time.

This places mainstream developmental psychology squarely in the temporal metaphysics represented by Newton. Nowhere is this more evident than in the explanations used to account for developmental phenomena. Consider a noted developmentalist's characterization of his field's *linear* emphasis upon the past: "The assumption that the cumulation of early life experiences, *beginning* in infancy, is critical for and determinative of later development and behavior has been almost universally embraced by human development scholars."[3] Consider also the widely accepted notion that developmental processes are *continuous* across time: "Whether one is an extreme hereditarian, an environmentalist, a constitutionalist, or an orthodox psychoanalyst, he is not likely to anticipate major life changes in personality after the first few years of life."[4]

Surely there can be no doubt that mainstream conceptions of development rely to a significant degree upon Newtonian assump-

31

tions of time and explanation. What is perhaps surprising is the number of *non*-Newtonian developmentalists. Although conceptions of time (per se) are rarely discussed in this literature,[5] Newton's temporal framework for explanation seems to have been increasingly challenged. Somehow, the utter dominance of a Newtonian framework for explanation, both for our culture and for developmental psychology, does not deter several "rogue" investigators from conceptualizing anomalies that challenge the very core of traditional developmental assumptions. This challenge could be easily dismissed as the musings of a few crackpots endemic to any discipline. But in this particular case, the rogues include some of the most stellar researchers in the field.

Perhaps it is *because* developmentalists have not had the option of overlooking issues related to time that they can have such dissenting opinions. Unlike other fields within psychology, they have not had the luxury of ignoring temporal factors. Maybe it is their field's dependence upon time that has caused some researchers to pause and reflect upon theories that contravene conventional explanation. At any rate, it is a mistake to assume that the field of developmental psychology depends upon certain temporal assumptions for its existence. This field, like so many other specialties within psychology, assumes a predominantly Newtonian framework for its explanations, but this has not precluded the development of significant anomalies.

To better understand this, the chapter begins with a brief history of modern developmental psychology. As it happens, the rise of absolute time is a primary factor in the rise of developmental psychology. The dominance of this approach to explanation has not, however, precluded other approaches entirely. Other historical notions of explanation are described which infiltrated developmental research and set the stage for significant anomalies. The state of modern developmental psychology is then examined. Its reliance upon Newton's five temporal characteristics of explanation is demonstrated, and a brief description of some "rogue" or "anomalous" views and findings is given.

The History of Time and Developmental Studies

The formal study of development is a strikingly recent phenomenon in the history of the world. Indeed, most scholars suggest that the actual discussion of development—as a topic in its own right—

began in the late seventeenth and early eighteenth centuries.[6] Thinkers who preceded this period were, of course, highly influential to current styles of developmental explanation. Nevertheless, most ancient thinkers had relatively little to say about children or their development across time.[7] This is probably surprising to most people in our modern era, because a person's development seems so obviously important nowadays. This was not always the case, however, and three factors seem to be the main reasons for this: a lack of absolute time conceptions, a high infant mortality rate, and little social mobility.

As discussed in Chapter 1, a Newtonian approach to time has not always been the dominant organization of world events. Currently, Western culture takes it for granted that all persons are continuous with their linear past, and thus the sequence of these linear events should be studied across time. Persons are thought to be inextricably linked to themselves in childhood, if not totally determined by their childhood experiences. Yet before the concept of absolute time developed in the sixteenth and seventeenth centuries, most people neither cared about time nor saw it in the continuous, linear framework represented by Newton. Humans, of course, have long had the mental ability to order their memories in "earlier" and "later" categories. However, the meanings of these categories have varied markedly across cultures and languages.

Few ancient cultures, if any, viewed these meanings from the absolute perspective of Newton and contemporary Western culture. Most societies were indifferent to consistent, absolute measures, let alone linear metaphors. People rarely troubled to remember their own ages, or the "date" of important events.[8] If time was noticed at all, it was governed by the person's *experience* of events rather than the person's experience of events being governed by time, as it is currently portrayed today. Work was over when the workers were finished or too tired, not when a timepiece gave its permission. Food was eaten when convenient or when people were hungry, rather than when allowed by the clock.[9]

This indifference to time (by present standards) was particularly detrimental to the study of development because events were not seen as linear or continuous with one another. No objective entity (such as linear time) existed to connect events, especially over long periods such as a life span. Childhood, in this sense, was not considered to be continuous with adulthood. Adult attitudes, emotions, and behaviors were thought to be qualitatively different and thus not "on a line" with one's childhood. Many events of one's life were marked, to be sure, but these events were not viewed as inevitably causal or continuous with one another as they are today. The first themes of temporal continuity were actually *cyclical* in nature. Using heavenly bodies as cyclical guides, people experienced events as constantly repeating themselves and began connecting them in a continuous manner, though the continuities were closed rings of recurrence.[10]

The notions of "earlier" and "later" were important to cyclical thinkers, but such notions differed greatly from modern linear conceptions. Our Western tendency is to assume that temporal concepts like earlier and later make sense only from a linear perspective. Earlier events are thought to have primacy over later events on a time line, with both sets of events being unique and nonrepeatable. However, as Aristotle once put it, "time itself is thought to be a circle."[11] When events occur on a circle, the meanings of earlier and later refer more to the order, rather than the primacy of events. No primacy is discernible because no event can "absolutely" precede another event.

For example, John's birth is "later" than George's harvest from a short-term perspective and considering only a small portion of the circle. However, John's birth is also "earlier" (on the circle) than the cyclical recurrence of George's harvest from a long-term perspective and considering the entire circle. Which event has (linear) primacy? No such distinction can be made, even in the order of one's *life* events. The only information available is the order of events relative to other events, not relative to some independent (and absolute) line of time.

The upshot is that the rise of absolute time facilitated, if not engendered, current interest in development. Developmental psychology is to some degree the result of a change in the predominant manner that events of the world are organized. Before this change, an event was considered and marked more in relation to other concurrent events—events happening at approximately the same time. The *new* organization of the world (of which Newton's writings are representative) related events to those that preceded or followed them. In fact, relations across time were considered the cornerstone of modern science because they were likely to be chains of causation or "determinants" of one another. These linear views ultimately led to the notions of progress and maturity that are so endemic to modern developmental psychology.

Health and socioeconomic factors also contributed significantly to the new awareness of development in the seventeenth century. A decreasing infant mortality rate was essential to any interest in development. As long as this rate approached four in ten children, the nuances of development were not given much emphasis.[12] The greatest intellectual energies were devoted to the greatest need— helping infants and younger children to *survive*. With advances in medicine, however, interests turned to less vital pursuits. Relations across time became significant—with the help of the new linear perspective—and the growth of children became an object of study.

Changing socioeconomic factors also fueled this study. The fuel in this case was the greater social mobility of the European citizenry during the seventeenth and eighteenth centuries. Before this period, the social position of parents determined the future of their children. Unlike the socioeconomic flexibility of today, hard work and intellectual skills counted little toward improving one's lot in life. If one was born a prince or a pauper, then one remained a prince or a pauper. By the end of the seventeenth century, how-ever, the industrialization of economies made social mobility a real possibility.[13] Most people experienced mixed emotions about this greater freedom: *joy* because persons were, for perhaps the first time, able to better themselves, and *fear* because a person's even-tual social and financial status was not assured. In other words, the rewards and worries of modern parenting were finally emerg-ing, and the call for a greater understanding of development was beginning to be heard.

A number of pivotal thinkers heeded this call during the pe-riod. Foremost among them was John Locke. Locke cogently ar-gued that the early experiences of children provided the foundation for adult behaviors and attitudes. His conception of the mind gave early childhood experiences a *tabula rasa,* or "blank slate," upon which to etch their (linear) influence. It was no coincidence that Locke was a close friend and, in his own words, an "underlaborer" of Newton.[14] In fact, Locke and his followers wished to be "Newtons of the mind,"[15] tracing the chronology of mental processes, just as Newton traced the chronology of physical processes. Although each man addressed different aspects of the developmental enterprise, each complemented the other's work. Newton supplied the meta-physics of development—the physical standard and justification for relations across a span of time. Locke supplied the epistemology of development—applying Newton's assertions to humans by showing how early events in childhood could be continuous with later events in adulthood.

Fully a century later, another pivotal thinker—Charles Dar-win—made his own important contribution to developmental psychology. His evolutionary perspective gave the study of development a scientific legitimacy that the philosopher Locke could not provide.[16] Although Newton had supplied the "physical" justification for relations across time, Darwin solidified this type of explanation as *the* explanation of science. Just as species evolve into other species, so children evolve into adults. This analogy was especially helpful in overcoming the older view that developmental events were independent of one's adulthood. After all, individual species were once considered to be separate and independent. Now that species had been shown to be connected to one another across time, it was but a small step to connect the experiences of childhood with those of adulthood.

Darwin actually made this connection himself after the birth of his daughter, Doddy, and wrote one of the first baby biographies as a result. Again, the analogies with his concept of evolution are direct. His theory of evolution stemmed from his detailed observations while on the famous voyage of the *Beagle*. Once he returned from this voyage, Darwin applied the same observational skills and conceptual perspective to his firstborn child. He recorded her life events in the same linear and continuous style as his theory of evolution: describing the *slow accumulation* of visual skills, and the *gradual* coordination of eyes and hands.[17] In fact, he briefly outlined many of the questions modern developmental psychologists have since attempted to address.

The net effect was that Darwin not only endowed the study of development with legitimacy, but he also provided developmentalists with a ready methodology and theoretical perspective. The linear and continuous assumptions of Newton had become reified as the way the world was organized. Although many evolutionary (and developmental) investigators have since questioned these continuities, an entire research program and conceptual outlook had been launched. Freud's theorizing, of course, epitomized much of this conceptual outlook. From his perspective, the past holds the key to psychological problems in the present, and the farther back we look into the past, the more basic the answers to the present are. Relations *across* time had become more important than relations *within* a particular period of time. Children were thought to be more completely understood through their past experiences than through their present experiences.

Actually, this brief historical sketch omits a significant, but in some ways separate, strand of current developmental thought—structuralism. This strand is, in fact, responsible for many of the anomalies now facing the Newtonian paradigm, though in most cases structuralism has been co-opted into the linear paradigm. Rousseau, albeit not a structuralist himself, influenced this school of thought extensively.[18] Unlike Locke, Rousseau downplayed the role of early experiences and the continuity of life events in development. He distrusted the linear notion that parents "shape" or "mold" (in the sense of "determine") their children's attitudes and behaviors. He believed instead that people possess an innate capacity for free will. He realized that such a capacity allows a child to act contrary to their early experiences. For this reason, he advocated that children guide their own development to some extent and that teachers understand how children construct their own experiences.

Although modern structuralists have de-emphasized Rousseau's notions of free will, they have kept his "constructivist" approach to knowledge as well as his emphasis upon the active role of children in their own learning. Piaget is an example of a modern structuralist. Similar to Rousseau, he distrusted purely linear and continuous conceptions. He relied instead upon holistic cognitive structures that the child employs to organize and construct the world. These holistic structures are not "molded" from early experiences but are endowed genetically. Because they are "wholes," they cannot be changed developmentally without some discontinuity (qualitative change) with past cognitive structures. A change in even a small piece of a structural whole can result in the meaning of the whole—the relations between the parts—shifting abruptly. This has led structuralists to value not only discontinuities in development, but also factors that occur concurrently with development (i.e., the present relation among the parts of the whole).

This structural strand of developmental history, though separate in its origin, has been intertwined with the linear strand more recently. Freud is partly responsible for this integration, being both linear and structural himself as a theorist.[19] The result is that discontinuous phenomena and relationships between concurrent events appear to be valued but are typically placed on a time line and themselves made continuous. Several modern theorists, for example, postulate a series of "holistic" stages in development that are continuous with one another. Each stage is derived from the one preceding it, and children are clearly continuous with their pasts.[20] The upshot is that a *mixed* temporal model of explanation is now the legacy of the structural school of thought. Let us therefore examine this legacy in modern developmental explanations.

The Newtonian Framework and Developmental Stages

Modern developmental psychology has come a long way from its conceptual origins. It now spans many varied topics and occupies an army of investigators. Still, the original philosophical justification and socioeconomic need for knowledge about development have remained intact. Its philosophical justification is the Newtonian (and Darwinian) paradigm that currently engulfs psychology. Relations across time are the sine qua non of modern understanding. Its primary socioeconomic need has not really changed since the first parents worried about the eventual social position of their children. Because "the child is father of the man," as Wordsworth observed, early developmental influences are viewed as crucial to virtually every aspect of adulthood, from physical characteristics to psychological health.

This begins to reveal the essential Newtonian character of developmental psychology. Although some "mix" of structural and Newtonian frameworks is common, it seems clear that the Newtonian element of this mix is the more dominant in mainstream theorizing. Of course, this varies from theorist to theorist. The present survey is not intended as a characterization of individual developmental psychologists. Rather, it is intended as a clarification of the various elements of this mix so that analyses of specific theorizing can be conducted. Each element is thus described in turn, with the Newtonian elements of developmental theorizing delineated first and the more anomalous elements of its structural heritage outlined second. It should be recognized, however, that specific theorists and researchers are likely to vary somewhere in between these two poles of developmental thought.

Certainly, all five characteristics of Newton's temporal framework for explanation permeate this field of endeavor (see Chapter 1). Recall that for Newton all processes occurred *in* absolute time and thus were to be explained in objective, reductive, linear, continuous, and universal terms. The first two characteristics are implicated in the very framework of the field. The objectivity of developmental phenomena and the reductionism of development itself are foundational to the formal and informal nature of the field. As noted above, most developmentalists assume an *objective* standard of time in their investigations and theories. Developmental events are thought to take place *in* time, and time is employed to measure and compare such events. Time is not considered a subjective entity that changes in relation to the observer or the phenomena being observed. Indeed, it was the historic movement away from such relativistic conceptions that gave the study of development its original impetus and scientific legitimacy.

Reductionism is more an informal feature of this field. Although the subject of study is development—human growth across time—many researchers reduce this study to a portion of development (e.g., newborns). These researchers attempt to explain and describe the physical and psychological conditions of a selected period of life. What are newborns capable of doing? What should a parent expect of an adolescent? Indeed, many psychologists spend their entire research careers in the study of only one age group.

Although the need for such research is unquestioned, the relations among investigations at various ages are often unclear. That is, concentration on each phase of life does not necessarily imply knowledge of the relationship *among* the phases. The forest can be overlooked for the trees, and several "life-span" developmentalists have noted this problem.[21] As Schaffer puts it,

> A great deal of information has been amassed now as to the nature of children's behavior at different ages and stages. . . . What is missing, however, is knowledge regarding the mechanisms of development. We may know how children behave at point A and how they behave at point B, but we have little idea as to how they get from A to B. The "why" of psychological development, i.e., the mechanisms responsible for developmental change, remains elusive.[22]

This problem is not totally unexpected, given the primarily Newtonian framework of developmental psychology. Processes that are assumed to occur across time, such as development, are inevitably reduced by the modern scientist to their parts as they occur across time. Because the process takes place through time, only a piece of the process is thought to be observable at any moment in time. *The overall structure of development is never directly accessible.* Researchers can enlist the aid of recording devices to "observe" an across-time process all at once, but this is an *in*direct observation.[23] The developmental process-as-a-whole is not itself being observed. Each portion of the overall structure must be "photographed" as it is separated by linear time, then these recordings must be placed together simultaneously to begin to understand the gestalt of development-as-a-whole.

The consequence is that developmental psychologists are relegated to a reduction of their subject matter. The phenomenon of development is reduced either to a portion of the process at any one moment of its unfolding (via direct observation) or to the distortions inherent in assessment devices that "record" all the moments and thus enable the overall structure to be examined simultaneously (via recordings). In either case, investigations of these across-time relationships are difficult. Indeed, this difficulty is a major reason that investigators specialize in one or more developmental periods; specialization seems to simplify the research. Unfortunately, it also makes it likely that part of the overall structure of development is being studied without a full understanding of that part's relationship (or meaning) to the whole.

Frequently, the justification for specialization—whether implicit or explicit—is that linear time itself has supplied the relations between parts of the whole. A full understanding of the life span is presumed to be unnecessary, because each developmental period is related to other periods through their linkage in linear time. Most specialists would readily admit to some unknown relations between periods of development. Still, many relations are *already* thought to be implied by their relationships along the line of time, e.g., early periods are foundational to later periods. Even if the "betweenesses" of developmental phases are not fully understood, specialists can supposedly take comfort in the fact that their findings are ipso facto *continuous* with any periods before and after, *linearly* causal to subsequent periods, and *universally* applicable to like-age children in similar circumstances.

Consider how research into the prenatal environment often presumes a *linear* significance to later stages. Indeed, many psychologists believe that the prenatal stage is "the single most important developmental period of our lives."[24] This emphasis on the first stage of growth is expected from a Newtonian framework because the earliest is always the most fundamental. Maternal stress, for instance, is considered to cause unwanted effects in later development.[25] The past is essentially considered immutable, and since time is *continuous*, this past is thought to have its inevitable "effect." Even when studies question the cogency of such prenatal explanations,[26] many researchers feel that convincing studies are the only thing standing in the way of validation.[27] Because events are arrayed along the same line of time, it is assumed that there must be *some* connection.

The newborn stage of development garners similar treatment. Newborns are quite sensitive to interpersonal relationships as well as surprisingly adept at imitating adults.[28] The quality of these early interpersonal experiences is thus considered to have linear significance to later development. For example, Klaus and Kennel assert that the bond between parent and newborn is directly correlated to their future relationship in life.[29] These researchers specifically define this bond as a unique relationship that endures through time and offer research spanning two decades in documentation. The long-term effects of newborn events are controversial, and other researchers have expressed some skepticism about the conclusions of Klaus and Kennel.[30] Nonetheless, many developmental explanations consider early nurturing events to be consistent with (and predictive of) many other factors later in life.[31]

Similar conceptualizations can be found at virtually any stage of development, from the infant to the elderly adult. Developmentalists routinely employ linear and continuous explanations, and their study of each stage derives a large portion of its significance from its assumed effect upon subsequent stages. Implicit, of course, is the *universality* of developmental psychology. Recall from Chapter 1 that Newton's linear and continuous explanations are considered valid regardless of the particular time in which they are observed. The so-called laws of nature are true for all times and situations. Similarly, most developmentalists search for laws of development that transcend (or generalize across) time and the particular children under study.

Let us turn more directly to these across-time theories. Unlike the description of a particular developmental period, these theories attempt to connect across-time events explicitly. Included here are physical and motor development, cognitive development, personality development, and social development. A thorough review of these is impossible in this context. Modern developmental psychology is too complex and intricate for our limited space.[32] Suffice it to say, however, that Newtonian explanations abound. We have already discussed how objectivity and reductionism are implicit to the very framework of the field, but continuity, linearity, and universality also hold commanding conceptual positions. To illustrate, let us briefly examine explanations in social development.

Newtonian Explanations of Social Development

Social development is thought to be the process by which a child learns to relate to other people. Most relevant researchers and theorists view this process as *linear*, in the sense of giving primacy to the past, *continuous*, in the sense of occurring consistently across time, and *universal*, in the sense of applying to all or most children. At points, structuralist influences appear to jeopardize the domination of these Newtonian explanations. Similar to the rest of developmental psychology, the historic influences of thinkers such as Rousseau and Piaget[33] are blended into this particular branch of developmental psychology. As we shall see, however, these influences are frequently more apparent than real, and ultimately get overruled by Newtonian factors in most mainstream theorizing.

Again, early stages of development are typically considered the most important to later social skills. The "secure attachment" of a mother to her baby, for example, is thought to have many later consequences in time. Such children are viewed as more at ease with other children, more interested in exploring new play situations, and more enthusiastic and persistent when presented with new tasks.[34] Some psychologists have contended that infants left in day-care settings are less securely attached to their mothers and more likely to assert themselves than children cared for at home.[35] Although the quality of day-care appears to be an important factor in such findings, this does not change the assumption of day-care's linear significance to later life.[36]

Of course, such linear theorizing places quite a burden upon parents. Parents are the first and most constant factor in any child's environment. Therefore, many developmental psychologists consider parenting one of the most influential and lasting effects upon a child's social development.[37] In fact, theorists have suggested that some approaches to child-rearing have substantial advantages over other approaches.[38] Baumrind, for example, holds that authoritarian parents (parents who require rigid obedience) are likely to produce children who are withdrawn and distrustful.[39] Other developmentalists have attempted to demonstrate that parental restrictiveness leads to dependence and submissiveness in children and that extremes of either restrictiveness or permissiveness result in later social problems.[40]

Other issues within social development—e.g., social cognition or sex roles—take on similar conceptualizations. Many cognitive developmentalists, for instance, believe that there are predictable changes in a child's "social cognition," depending upon the early experiences and cognitive stage of the child.[41] Sex roles are similarly considered the result of cultural influences from the past.[42] But like other realms of developmental psychology, certain strands of structuralism are mixed into these predominantly linear and continuous interpretations of social development. Some interpretations of social cognition, for instance, emphasize "constructive" and "discontinuous" structural influences. These can seem at variance with strict Newtonian characteristics.

For example, constructivist explanations appear to violate the linearity of time. Because children are thought to "construct" their knowledge of social relations in the present, the traditional (linear) primacy of the past may seem in jeopardy. A major question arises, however, when the developmentalist is asked about the cause of the particular constructive act. The developmentalist is faced with accounting for some new or original element in the constructive act, when the only explanatory models that are available portray everything as an effect of some previous cause. Clearly, in most mainstream theorizing linear time and causation continue to hold sway. That is, questions about the origins of the construction lead most modern developmentalists to some prior determinant, either biological or experiential.[43] In this sense, "constructions" are really a product of factors from the past, and remain within the Newtonian framework for explanation. (For a similar analysis in cognitive psychology, see Chapter 5.)

This is also true of many discontinuous stage theories. Many social cognitivists, for instance, discuss "qualitative change" and "discontinuity" between stages of social development.[44] Children are thought to move from one qualitatively different stage to another, seemingly in contradiction to Newtonian notions of continuity. As Tomlinson-Keasey notes, however, such explanations have not been popular with American psychologists.[45] Indeed, a complete explanation of such discontinuities often reveals a Newtonian spin. For example, discontinuous changes are defined as "spurts" of continuous development where there are points of linear time (and states) *in between* stages.[46] This means that such discontinuities are merely faster-paced *continuities* rather than truly abrupt shifts in developmental status from one instant to the next (and no time "in between"). Changes in particular stages, albeit accelerated, are considered to occur smoothly through "increasing differentiation" or "a continuous process of change."[47]

The issue of qualitative change also has important implications for the *universality* of developmental processes. If one stage is qualitatively different from another, then the "laws" that govern one stage would not necessarily govern another stage.[48] Each would be qualitatively distinct from the other. An understanding of developmental processes *within* a particular stage would not necessarily apply to processes at another stage, making the processes unique to the specific temporal context rather than universal to all temporal contexts. If, on the other hand, the transition from one stage to the next is continuous, the dividing line between stages would be blurred. A *continuum* of basic properties—varying only in quantity (and not quality)—would be implied. Moreover, any explanation about the basic qualities of a particular "stage" would be presumed to have universal applicability to any other stage—only the quantity of those qualities would be in question (e.g., children are more concrete thinkers). In this sense, most developmental principles are viewed as universal, and most researchers are attempting to find static laws that transcend particular developmental contexts.

To conclude, then, the Newtonian metaphysic continues to permeate modern developmental psychology. Newton and his intellectual compatriots, Locke and Darwin, have helped spawn a field of endeavor that examines all aspects of human behavior across linear time. Although many developmentalists are not directly concerned with across-time explanations, the significance of study for even a particular developmental period is often derived from its presumed consequences later in time. Further, investigations of across-time topics, such as social development, show clear Newtonian proclivities, despite the structuralist legacy of constructivism and discontinuity.

Anomalies to the Newtonian Paradigm for Development

The question remains whether developmental psychology has any anomalies to its Newtonian paradigm. As noted in our brief historical survey, this field clearly has the seeds of such anomalous thinking, given the influence of structuralism. From the above discussion, however, Newtonian assumptions appear to override structural influences in theorizing. Even when constructive and discontinuous explanations are attempted, the primacy and consistency of the linear past are ultimately affirmed. Are there *genuine* anomalies to the Newtonian framework? Despite a general dependence on a linear theorizing, there are, most assuredly, several intriguing instances of nonlinear explanation in developmental psychology.

Perhaps the most prominent is what developmental researchers have termed *discontinuities* in development. Unlike our discussion of stage theory above, this particular form of discontinuity is not predictable and not governed by early experiences or continuous change.[49] This anomalous form of discontinuity entails an almost puzzling *lack* of predictability across time. Whereas continuities refer to connectedness in development—linking early behavior to later behavior—discontinuities in this context allude to the lack of any such link.[50] Developmentalists who assert the existence of this sort of discontinuity are verging on stark Newtonian heresy. For Newton, the chief property of the world was its predictability. Because events of the world take place in and across absolute time and time itself flows in a consistent and continuous manner, the present and future are *always*, in principle, predictable from the past. To assert that there is little or no *fundamental* predictability between a person's early and later behaviors is to place in jeopardy the explanatory principles of linearity and continuity, if not universality and reductionism as well.

Nevertheless, many developmentalists continue to confirm such discontinuous relationships. Indeed, this confirmation has occurred *in spite of* their expectations to the contrary. The linear paradigm of psychology led most researchers to anticipate the "indelible effects" of early experience.[51] Still, the number of instances in which these indelible effects did not occur repeatedly surprised these researchers. In their book, *Continuities and Discontinuities in Development*, Emde and Harmon put it this way:

> Developmental investigators engaged in longitudinal study were disappointed in finding little predictability from infancy to later ages. This was true for behavior related to cognition (see McCall, 1979; Kagan, 1984) as well as for behavior presumably related to temperment (Plomin, 1983). During this time clinicians, who had assumed indelible effects from early experience, were repeatedly surprised by well-documented instances of resiliency following major infantile deficit and trauma (for review see Clarke and Clarke, 1977; Kagan, Kearsley, and Zelazo, 1978; Emde, 1981).[52]

Linear theorists assume that traumas in early childhood have long-lasting and potentially devastating effects, far into the future. On the contrary, replies Jerome Kagan, "You can predict very little from experiences at ages 0 to 3, even when they include a number of traumatic events."[53] Indeed, complete recovery from such traumas—without intervention—is not uncommon. Consider a longitudinal study of over 200 Korean girls. Results showed that the degree of malnutrition at the time of their adoption (2 to 3 years of age) had little to do with any subsequent variables six years later. In fact, the average IQ of the most severely malnourished group was forty points higher than expected.[54] Studies of normal children with relatively stable environments have revealed similar discontinuities. Variations in psychological qualities during the first three years of life were *not* predictive of variations in culturally significant and age-appropriate characteristics five, ten, or twenty years later.[55]

Faced with such discontinuous findings, the first thought of investigators was that their methods required refining. The problem of discontinuity *had* to be methodological because a lack of continuity and predictability was considered impossible. As Kagan notes, "most Western scholars presume a structural connection between all major phases of development with no breaks in the story line."[56] However, the problem got worse when methods were refined. Instead of the expected discoveries of reliable correlations between early and later behaviors, the *absence* of such linear continuities was increasingly confirmed. Emde and Harmon put it this way: "expectations about continuities received a jolt from two decades of *incontrovertible* research findings."[57]

Other developmental researchers seem to concur. Rutter sums up a large portion of this literature rather succinctly: Correlations "from the infancy period to maturity [are] near-zero."[58] Jerome Kagan also notes the lack of linear primacy: "The early form [of development] did not determine the new one; the new entity was not contingent on the earlier structure or process."[59] Novel systems of development seem to emerge not only in early childhood but also in later stages of development, including old age.[60] Many researchers contend that "humans have a capacity for [qualitative] change across the entire life-span."[61] This has led Rutter to claim that some "behavior can be understood and accounted for solely in terms of contemporary or ongoing events."[62] In other words, the focus of developmental psychology upon the past is at least partly misdirected. If persons can change dramatically at any time during their life span, the present has as much to do with behavior as the past.

Does this mean that there are no *continuities*? Although this question has not been completely answered yet, most developmentalists view the findings so far as suggesting *both* continuities and discontinuities.[63] Kagan, for example, has shown how most longitudinal studies have failed to find continuities in popular molar behaviors, such as irritability, activity level, and dependence on parents.[64] Nonetheless, he has also found a tendency for "behavioral inhibition" to remain fairly stable in later life. That is, small children who withdraw from unfamiliar or threatening situations continue to exhibit similar behaviors in school. In fact, these same children choose vocations as adults that appear to minimize competitiveness and traditionally masculine tasks.[65] Such findings exemplify the current mix of continuities and discontinuities presently being studied.[66]

One temptation is to assume that this is merely an empirical issue—that the researchers will soon ferret out what's what and that will be the end of that. It is important to realize, however, there are many aspects of this controversy that are *not* empirical in nature. Underlying Newtonian biases reveal themselves in several significant ways. First, such biases lead researchers to *assume* the existence of continuities in developmental processes. Thompson and Lamb illustrate this nicely: "the question is not whether continuity exists, but under what conditions it occurs."[67] Second, these same biases lead investigators to assume that discontinuities do *not* exist. Emde and Harmon, for example, have had to warn investigators not to view discontinuist data as "negative" or "null."[68] The implication is that many investigators are likely to ignore such data. They may consider it "error variance" or the nonconfirmation of continuist findings.

Current psychological method is also suffused with temporal factors favoring continuities. As Chapter 4 demonstrates, most forms of modern experimental design are based upon the notion that important scientific processes are continuous and linear in nature. Consequently, conventional research designs cannot detect discontinuities without difficulty. For instance, how do researchers know when they have sufficient experimental power to detect the *absence* of continuous phenomena? Lipsitt seems to reflect the theoretical prejudice of many investigators when he says: "Apparent non-continuities may be instances of continuities not yet fully revealed" or "continuities not yet sufficiently investigated."[69] Moreover, modern methods lend themselves to linear interpretations of data. Because independent variables supposedly precede dependent variables (and causes supposedly precede effects), it is only natural to assume that the very nature of any data is linear and continuous.

Rutter, Quinton, and Liddle cite an interesting example of a methodological practice that favors the finding of continuities. They note that the continuity of "looking backward" in research is not the same as the continuity of "looking forward."[70] For instance,

Robins and others have shown that most sociopathic adults are delinquents as children (looking backward).[71] At first blush, this seems like an obvious continuity. This would appear to imply (looking forward) that a delinquent child is very likely to grow up to be a sociopathic adult. The fact is that only a *very small minority* of delinquent juveniles become sociopathic.[72] That is, the linear type of conceptualization in no way accounts for the vast majority of delinquents who reach adult life without becoming sociopathic.[73] What might be taken as an obvious continuity may actually be a discontinuity of sorts. Looking forward, the large proportion of delinquents who go on to become highly productive and caring members of society can be viewed as a discontinuity. Metaphysical biases may lead some researchers to see what they expect, and still more researchers to stop short of a full understanding of the data they have gathered.

Kagan calls attention to the ease with which continuist theories are constructed. He feels that many continuities have been based upon "superficial similarities between one aspect of the behavior of an infant and a single aspect of an action noted in older children or adults."[74] Freud, of course, was famous for this sort of connection-making, and many of his "continuities" may seem farfetched to us now. In his *Three Essays on Sexuality*, for example, he asserted a similarity between a nursing infant and adult sexual intercourse:

> The erectile nipple corresponds to the erectile penis: the eager watery mouth of the infant to the moist and throbbing vagina, the vitally albuminous milk to the vitally albuminous semen. The complete mutual satisfaction, physical and psychic, of mother and child in the transfer from one to the other of a precious organized fluid, is the one true physiological analogy to the relationship of a man and woman at the climax of the sexual act.[75]

Although few contemporary developmentalists would subscribe to these particular connections, Kagan feels that many current continuities may meet the same notorious fate as this "continuity" from Freud. Such an example certainly demonstrates how easily such continuities can be formulated.

Indeed, many investigators believe they are literally "invented," rather than discovered.[76] This, of course, challenges the *objectivity* of absolute time assumptions. Continuity (or discontinuity) is not "out there" in the world or the data; continuity is a construction of what the data mean to the investigator.[77] Empirical findings are still relevant, but data in and of themselves do not have meanings. Empirical findings require *interpretation*, and as noted about delinquency, one investigator's continuity is another investigator's discontinuity. Each interpretation is hypothesized from a theory that itself depends upon metaphysical assumptions. The assumptions themselves are not objective. They are intersubjective (shared) meanings with a power comparable to the data in deciding "the findings." This would mean that the recognition of one's metaphysical biases—in this case Newtonian temporal biases—would be crucial to an understanding of one's data.

Recent empirical and theoretical articles have provoked a more critical attitude toward continuities. Explanations based upon Newtonian features such as continuity and objectivity are held more skeptically. Even in animal and biological research, these assumptions have been challenged. Baldwin and Baldwin, for example, have noted that the early play behavior of squirrel monkeys was believed to be necessary for normal social behavior as adults.[78] Still, they have observed that the absence of such play behavior in many squirrel monkeys has not affected the appearance of normal social behaviors. This suggests that juvenile play does not significantly contribute to adult social behaviors. Some biologists have even proposed that each life stage has structures that are wholly temporary. These structures are important for maximal adaptation during that period, but they are no longer needed when the next stage is reached.[79]

When such conceptions are applied to the human species, they suggest that aspects of early development are temporary adaptations which have little to do with later development. Some aspects may disappear with development and not be connected with future qualities, even if they appear similar. In this regard, Kagan proposes the following as temporary developmental adaptations: the fear that infants have of strangers, the single-word speech of 18-month-olds, and the absolute definition of right and wrong typically held by 3-year-olds.[80] Consider, for example, the traditional developmental assumption that children must first pass through a stage of speaking single words before they can speak multiword utterances. Kagan contends that the universality of this intervening stage is *our* invention. Some children undoubtedly pass through this stage, but many children do not, and there is no consequence in either case for future speaking skills. Evidence of this sort challenges two Newtonian axioms: the primacy of the past (linearity) and the necessity of intervening stages (continuity).

Non-Newtonian explanations, such as temporary adaptation, are vital to the eventual resolution of the issue. These explanations postulate the positive existence of discontinuities. As Kuhn teaches, scientists have to conceptualize the real possibility of something existing before they can see (or "invent") that something in their data.[81] It is not sufficient for discontinuities to be defined negatively as the "absence of something" (namely, continuity). They must have a theoretical identity and explanation all their own to be real possibilities. Interestingly, preliminary steps toward this necessary theoretical identity have already begun.[82]

Unfortunately, these steps are greatly hindered by the unrecognized dominance of the Newtonian framework in developmental psychology. This framework leads only to continuous and linear theories. The presence of discontinuities not only violates the expectations of continuity and linearity but also the expectations of universality. Linearity, of course, is violated because some early developmental processes have no influence whatsoever upon later developmental processes. Their presence or absence may affect current functioning but not later functioning. Newton's notion of continuity is also violated because many early and later developmental processes are not consistent with one another. Moreover, continuist notions of change are violated since some processes literally exist at one point in time and do not exist at another point in time. Thus, cumulative and gradual notions of change are not the only conceptual possibilities.[83]

Discontinuities also defy the universality of most developmental explanations. If there is no consistency of a particular developmental process across time, then knowledge about that process would be unique to the particular temporal context. That is, knowledge about the process would only apply to the "moment" (or developmental period) in which it is occurring. As a discontinuity, it would not be predictive of the person's future. The manner in which the child went through the process (good or bad) would have no import on future developmental issues. Furthermore, some children might omit the process altogether with no ill effects. This would imply that the developmentalist cannot even say that the temporary adaptational process is universal across people.

The upshot is that the stability necessary for universality—the conventional notion of scientific "law"—would not exist, either across time or across people. The only knowledge available for such phenomena is knowledge that is applicable to the specific time and individual involved. As Sandra Scarr has correctly noted, this approach is quite compatible with a thoroughgoing structuralist (or constructivist) approach to knowledge.[84] The term "thoroughgoing" is used, because structuralist influences are often subjugated to the more dominant linear influences in the mainstream literature (as described above). Nonetheless, as this review of anomalies has demonstrated, the nonlinear side of developmental psychology sometimes also prevails.

Scarr reveals this nonlinear side in her discussion of universality. She boldly states that developmental conceptions are "products of their own times and spaces."[85] As we saw above, this was the ultimate implication of discontinuity research. Scarr, though, turns her attention to other time-based "inventions" of the researcher's mind as well as their practical consequences:

> Suppose we had considered alternative theories about early experience [linearity] and the permanence [continuity] of everyday traumas. We might have spent less time looking for far-flung sequelae of temporary maternal separations and more time looking at people's *current* lives.[86]

Part of her contention is that it is our theoretical emphasis on early experiences and its "permanence" across time—*not* the data of development itself—which has led to current explanations. She observes that our infatuation with early experiences (the primacy of the past) led to the Head Start program. She reasons, alternatively, that we could have easily viewed adolescence as more fundamental, and thus ended up with "Teen Start." This would have been equally compatible with developmental data and potentially just as beneficial. The major difference between these two practical consequences—Head versus Teen Start—was not the data but the theories (and assumptions) employed to interpret the data.

What governs the employment of a particular theory? Scarr feels that the culture—the time and place of theorizing—has a great deal to do with this. In the 1950s and 1960s, she notes, the nuclear family was viewed as the ideal. Children from "broken" homes were considered "at risk"—one of these risks being a son's decreased masculinity because his father was likely to be out of the home. With the advent of the women's movement, however, scientific assumptions of broken homes changed. Suddenly, we had alternative forms of the family and nontraditional households.[87] Now, of course, it is appropriate for a woman to work outside the home and for sons to avoid traditional masculinity. Now, the ideal may involve becoming a more nurturing, androgynous male.

Scarr's point is the context specificity of developmental knowledge. What is "known" is not an unchanging law across all time (or space) but rather an interpretation of what is going on, and this interpretation is always relative to a specific temporal context, such as culture. Truth, in this sense, is not some abstract, never-changing realm "behind" the constantly changing appearances of the world. Truth does not universalize across all times and all situations. Similar to the (anomalous) hermeneutic method, discussed in Chapter 4 below, knowledge cannot be based upon some correspondence between hypothesized and changeless truth. Knowledge must rely upon truths specific to a particular metaphysic, culture, developmental phase, and person.

At this point, only one of Newton's five characteristics of explanation has been omitted from our discussion of anomalies—reductionism. Recall that one of the problems inherent in the informal nature of the field is a focus upon the trees at the expense of the forest. The developmental process-as-a-whole is often reduced to its component parts through the study of specific periods of the life span. A great quantity of information has been amassed about children at specific ages, yet little is known about the relationship *among* these ages. The difficulty is that the processes responsible for this relationship probably remain "elusive," to use Schaffer's term,[88] as long as the underlying reductionism of Newton's temporal framework is left intact.

However, as noted in the initial presentation of this issue, a relatively new specialty within this field has arisen to champion the whole of development: life-span psychology. Interestingly, it is from this particular subspecialty that many of the foregoing anomalies have originated. Life-span developmentalists have conducted much of the research on discontinuities. In addition, those theorists who have been the most outspoken about the "invention" (rather than the objective reality) of linear theories and the dependence of such theories upon sociocultural (rather than universal) factors have seemed mainly to issue from this brand of developmental psychology.

This really should come as no surprise. All five Newtonian characteristics of explanation stem from one central assumption regarding time. If one characteristic is questioned or challenged, then all are likely to come under scrutiny. The fact that all the characteristics have now been substantively challenged points to the growing need for a reexamination of the basic premise which underlies them. Kagan and Scarr seem to be verging on this, yet the importance of time has not been clearly acknowledged. This is not to imply that all developmental phenomena require a reassessment of metaphysical assumptions. As already mentioned, many continuities are widely affirmed. Still, it is not known how even these continuities will fare under the new light of a metaphysic that does not consider absolute time to be axiomatic.

Conclusion

The obvious need for time conceptions in developmental psychology has led to a heavy dependence upon Newtonian assumptions. This is not to say that all developmentalists hold such conceptions. Indeed, it seems that those developmentalists who take into account the greatest "span" of time are those who are the least Newtonian (e.g., life-span psychologists). Nevertheless, developmental psychology is clearly dominated by Newtonian temporal characteristics of absolute time. Most developmentalists assume that some objective standard for time, aging, and chronological order exists outside the subject matter being studied.

This has led, in turn, to explanations of developmental data that reflect Newton's other temporal characteristics. Most developmentalists suppose, for example, that later behavioral patterns are consistent and *continuous* with early behavioral patterns. Although such consistency does not necessarily imply that the past causes the present, most developmentalists make this *linear* supposition as well. Early experiences are considered to hold primacy in the course of all later psychological processes. Moreover, these past-to-present relationships are viewed as *universal* principles, cutting across temporal contexts and providing justification for *reductionism* in the discipline. That is, developmental psychologists assume that the thorough investigation of one temporal context (e.g., that of two year olds) has an axiomatic (linear) relation to anything that follows it in time. Therefore, the relation between temporal contexts is thought to be known to an important degree a priori and remains underinvestigated.

This does not mean that such relations are *not* investigated. Life-span developmentalists, for example, see this as their main task. Although their scientific investigations typically follow a Newtonian framework as historically inherited (see Chapter 4), their interpretations of these investigations seem to have trended away from absolute temporal notions. For example, they openly embrace discontinuities in development, which are not seen as simply accelerated continuous changes in development but instances of nonconnectedness and unpredictability between developmental stages across time. Indeed, there is important evidence that some developmental stages are wholly temporary, without any linear influence on subsequent stages or behavior.

The temptation has been to assume that these findings are methodological quirks or hidden continuities. This is in itself a great testament to the power of absolute time in developmental psychology. Many developmental researchers choose to trust their metaphysical axioms over their own dramatic and repeated evidence to the contrary. Purely structuralist theories have been similarly distrusted. Newtonian approaches have essentially reigned triumphant, whether in theory or research. Nonetheless, the evidence of discontinuities and the influences of structuralism do not seem to be going away. The challenge, therefore, is to find a metaphysic that has the explanatory power to conceptualize all aspects of the developmental enterprise.

Chapter 3

PERSONALITY THEORY

Perhaps no field in psychology has greater conceptual diversity than personality theory. Leading texts typically contain theories advocating everything from materialism to spiritualism, determinism to freedom, and reductionism to holism. This is due, in part, to the wide range of thinkers who are included in such texts. Many theorists, for example, were trained outside of the discipline of psychology and are not therefore tied to its intellectual foundations. Personality theory has also traditionally occupied a dissident role in psychology.[1] Most theorists have defied the conventional methods and ideas of their times and openly assumed the role of provocateur.

One of the conventional ideas that has been challenged, though not always explicitly, is the temporal framework of Newton. Hall and Lindzey have attempted to characterize the major differences among theories in the beginning of their classic personality text.[2] Of the "substantive attributes" they offer for comparison, all five characteristics of Newtonian temporality are included in one form or another. They note that theorists differ on issues of "continuity and discontinuity," the "importance of early developmental experiences" (or linearity), their emphasis upon "general principles and laws" (or universality), the degree to which they affirm "holistic principles" (or reductionism), and the significance of an "objective frame of reference" (or objectivity).[3] In this sense, then, the conceptual diversity of personality theorists has clearly extended itself to time-related issues.

More recent commentators, however, do not seem to affirm this temporal diversity in the field. Many seem to advocate the

55

Newtonian paradigm in their general assumptions of personality or their critical appraisals of individual theories. Several recent texts, for example, incorporate the explanatory characteristics of absolute time into their general definition of personality. Maddi defines it as a "stable set of tendencies and characteristics . . . that have continuity in time,"[4] or as Phares puts it, "persist over time."[5] Ryckman gives the construct of personality a distinct linear flavor, when he includes "the person's unique learning history"[6] in his definition. Similarly, Ewen and Maddi both assume, respectively, that "early childhood"[7] and the "family setting in which one matured"[8] are necessary factors in any personality theory.

General definitions of this sort seem to imply that a "proper" personality theory must have linear and continuous attributes. If theorists do not view personality as relatively stable across time and heavily influenced by the past, then they are either not personality theorists or liable for criticism. "Developmental" explanations are considered necessary for any good theory,[9] and science is viewed as requiring a Newtonian explanation. Ross, for instance, says that all explanations in personality consist of proximate and ultimate causes, both of which precede their effects. He feels that "as a psychologist I accept the logic of determinism."[10] All scientists, Ryckman declares, believe that "behavior is caused by previous events."[11]

Such definitions and requirements may institutionalize a Newtonian framework in the field of personality. Despite a strong tradition of conceptual diversity, new theorists and researchers may find only linear and continuous explanations acceptable. This could affect personality theorizing in several significant ways. First, any theory that does not meet linear criteria may be overlooked or ignored, being too substandard for a truly scientific (i.e., Newtonian) field of personality. Second, a nonlinear theory can also be misconstrued or "linearized." That is, it may be interpreted as though its nonlinear constructs are ultimately explained by linear factors, such as the past. Finally, nonlinear theorists may be viewed as invoking enigmas that are not amenable to "scientific" explanation (e.g., Jung's synchronicity). In this case, personality theory commentators do not even attempt to make a nonlinear explanation intelligible to their readers, labeling the explanation "mysterious."[12]

The purpose of the present chapter is thus twofold: (1) address the potential institutionalization of Newtonian methods and ideas in the field, and (2) draw upon the lessons of those theorists who have both affirmed and denied psychology's Newtonian paradigm. The chapter begins with the historical roots of personality theory. Here, the contemporary battle over the subjectivity and objectivity of time is depicted through the philosophies of John Locke and Immanuel Kant. Freudian theory, perhaps the first great conception of personality, is then shown to have a mixed philosophical heritage. The vast majority of personality theorists, however, is revealed to follow Newton's lead in their explanations. Exceptions to this, notably the work of Kurt Lewin and Carl Jung, are then described along with some of the problems commentators have had in understanding and explaining these theorists.

The Conceptual Roots of Personality Theory

The roots of personality theory can be traced to earliest recorded history. Because personality theory is vitally concerned with human nature, any thinker who has commented upon our nature can be considered an intellectual ancestor of the modern personality theorist. A review of all these thinkers is obviously beyond the scope of this chapter. However, broad over-arching themes can be delineated. A modern personality theorist, Joseph Rychlak, has shown that the philosophies of John Locke and Immanuel Kant represent these broad themes in many important ways.[13] These philosophies are the modern extensions of long lines of empiricistic and rationalistic philosophies that preceded them (e.g., Aristotle and Plato). As Rychlak has demonstrated, most modern approaches to personality are variations upon these two philosophical themes.

For his own purposes, Rychlak tended to concentrate on the epistemological differences between Locke and Kant—differences in how each felt we know things. Locke felt that our knowledge originated from two sources: *sensation* and *reflection*. Sensations, such as heat, softness, and bitterness, come from objects that are independent of our minds. The human mind is a tabula rasa at birth, and experience etches representations of the external world onto this blank slate. Reflections, such as thinking and reasoning, result from an internal sense which our mind possesses when it attends to its own operations. Simple sensations enter the mind as "atoms" of experience. These are distinct from one another even though a single entity may have caused them (e.g., color in a moving object). Complex ideas are those that arise from the reflective operations of the mind as it combines and relates simple and discrete sensations.

Locke thus established the modern empiricist's dictum: Knowledge is derived from experience. Although his empiricism was perhaps not as radical as many of his students,[14] he set the clear precedent for empiricism when he insisted that all ideas originate in a separate and external reality.[15] Ideas are built up from the atoms of experience, and put together in a representation of the object as it occurs external to the organism. In this sense, Locke opposed the notion of innate ideas. Knowledge did not stem from inborn forms, as proposed by Plato. Knowledge also was not rooted in ideas that were presupposed by experience and independent of sensations. Knowledge was instead "real" because experience of the external world began and determined it.

Kant agreed with Locke that experience is necessary for our knowledge, but his view of that experience was radically different.[16] Kant distinguished between *phenomena* and *noumena*. Phenomena consist of our experiences of the world as they are naturally organized by the inherent actions of the mind. That is, our everyday experience is not an experience of the world as it "really" is; it is an experience of the world as it is structured by our mind. The world as it *really* is—external to our consciousness—is the noumenal realm of existence. We never have access to this realm because it is integrated with our mental structure before we ever "experience" it. For instance, our experience seems to reveal to us the causes and effects of our world as we encounter it each day. According to Kant, however, causation is part of the organization that our mind imposes upon the noumenal reality of our world.[17] Causes and effects are not *in* the external world (independent of our consciousness) but extended from the mind as it naturally colors events.

The only knowledge that we truly have, then, is knowledge of our phenomenal realm. We can never know the noumenal realm empirically because it is always combined with cognitive structure. Kant did postulate certain *categories of understanding* that lent structure to experience from birth. Knowing them enabled him to infer, in some crude way, what the noumenal realm might be like. For instance, objects in the noumenal realm must be uncaused. Because cause and effect relations are mentally extended to things, rather than inherent in them, things-as-they-are (the noumenal realm) would lack causal connection. People, Kant speculated, must be noumenally free of causal constraints, particularly if moral responsibility is to have any meaning.[18] In science too, causes and effects are not detectable with instruments or experimentation because they are not "out there" to be detected or measured. Causation *is* experienced by the scientist but only because he or she naturally organizes the data of science in this manner.[19]

In this sense, then, Locke championed the objective and real as the primary determinants of knowledge, and Kant championed the subjective[20] and mental as the primary determinants of knowledge. Both, however, were nature/nurture interactionists. That is, each held that the native qualities of the mind and the nurturing qualities of the environment were significant to knowledge. Nevertheless, each also asserted a dramatically different mode of interaction between these two qualities. Locke placed the environment primarily in charge, whereas Kant placed the mind primarily in charge. Locke also presumed more of a bottom-up directionality to knowledge, with concrete and simple sensations building up to the more abstract and complex ideas. Kant, on the other hand, presumed more of a top-down directionality. His abstract categories of understanding moved "downward" to frame and give organization to the more concrete aspects of experience.

Rychlak and others have cogently argued that past and present theories of personality are drawn from the epistemologies represented by these two great philosophies.[21] Those theories that have loosely followed the *behaviorist* tradition—e.g., Dollard and Miller, Skinner, Wolpe, Bandura—conceptualize the person in the spirit of Locke. Those theories that have loosely followed the *humanist* tradition—Rogers, Allport, Kelly, Binswanger, Boss—conceptualize the person in the spirit of Kant. And those theories that have loosely followed the *psychoanalytic* tradition—Freud, Adler, Jung, Horney—conceptualize the person with a mixture of Lockean and Kantian epistemological principles. Most of these epistemological distinctions are fairly well recognized by the modern personality theorist.

Objectivist and Subjectivist Models of Time[22]

What is not well recognized are the differences between Lockean and Kantian philosophies concerning linear time. These differences are fundamental not only to the philosophies themselves, but also to the personality theorists who have employed these philosophies in their concepts of human nature. As noted at the beginning of this chapter, the unrecognized Newtonian framework of this field has led many psychologists either to ignore theories that do not fit this framework or to linearize them with an interpretation that forces them into a Newtonian explanation. Exposing the temporal assumptions of these two basic philosophies should help prevent this as well as provide a better sense of the options available for personality theorizing.

Locke is representative of the *objectivist* view of linear time. Linear time is independent of consciousness, and we know of it through our experience. Locke's views are in some ways the epistemological extension of Newton's metaphysics (see Chapter 2). The philosophical similarity of the two friends is indicated by Locke's pronouncement that he was an "underlaborer of Newton."[23] Although Locke's epistemological interests kept him from explicitly asserting the absoluteness of time, he did appear to assume that the absolute existence of time had been demonstrated and that it possessed many of the properties which Newton claimed.[24] As Locke put it, "duration is but as it were the length of one straight line extended ad infinitum."[25] Linear time was for Locke a complex idea because it was a relating together of the succession of simple ideas received from the world. Linear time, then, exists and begins in the objective data of sensory experience.

Locke's philosophy of human nature reflects his views of linear time. Because all events must occur along the line of time, his empiricistic epistemology assumes that the external environment (via sensations) has an objective temporal priority over the internal operations of the mind. The environment is epistemologically primal, so the environment has to be temporally primal. Moreover, the mind itself is "built up" across time. Simple sensations are accumulated through time and gradually made into complex ideas later in time. The tabula rasa upon which these sensations are etched is a crude analogy to modern information storage conceptions. This style of theorizing allows the mind to preserve the past for *continuity* and later linear effects. In this manner, the mind is always an effect of previous etchings or "inputs," and one's behavior and personality are always consistent with, if not determined by, one's previous experiences.

Kant, on the other hand, is representative of the *subjectivist* view of linear time.[26] For him, time is a "subjective condition . . . a form of internal intuition" which a person naturally and unconsciously uses to structure reality.[27,28] Similar to the example above of causal relations being extended to events of the world, he believed that linear relations were also extended to events of the world. In this sense, Kant agreed with Newton that linear time was independent of the content of the noumenal realm. But he disagreed that linear time was an entity that existed outside the mind.[29] We are born into a mass of sensations (the noumenal realm), though we never truly know it. This is because we come equipped with a priori principles of organization, such as linear time, which allow us to sort the mass of sensations into linear sequences. In other words, the noumenal realm does not possess these linear characteristics. Only the phenomenal realm—the gestalt of the noumenal and mental—possesses linear relations.

Kant's philosophy of human nature also reflects his assumptions of time. Unlike Locke, Kant did not view human nature as built up across time. Innate categories of understanding are an unchangeable source of mental structure, and are thereby unaffected by events across time. Experiences, then, are conceptualized by categories that are themselves independent of linear time. Sensations do not simply come "in" to be added to the pile of yesterday's inputs. Sensations are continually being conceptualized at each instant of their perception, giving Kantian philosophy an "ever-present" style of knowing. Although past (phenomenal) experiences are not ignored, conditions of the present—including the present organizations of the mind and situation—have more to do with behaviors. Many Lockeans may seem to emphasize the present organizations of mind and situation, but their objectivist view of time means that these organizations are in linear sequence and themselves products of past determinants.

Less well known is the transcendent quality of Kant's philosophy. Humans not only escape linear determinism through the ever-presentness of the categories of understanding, humans also reconstrue their phenomenal knowledge of the past. Storage metaphors are less applicable to this epistemology. Just as the present is construed by the conceptualizer, so too, the past is construed (in the present) by the conceptualizer. Kant's *transcendental dialectic* permits the mind to go beyond what is already known, and explore possibilities not conceptualized in the past.[30] Recall that Kant held that the noumenal self was (by inference) uncaused, and so people are free of such linear constraints.[31] We can reason to and act upon patterns of behavior not encompassed in our (phenomenal) past experiences and thus be discontinuous with prior experiences across time.

If the original correlation between psychology's personality traditions and these philosophies hold true, then this additional information about their views of time offers us fresh insights. As demonstrated below, those in the behavioral tradition stress a

Lockean, objectivist view of linear time, and those in the humanistic tradition emphasize a Kantian, subjectivist view of linear time. Perhaps more controversial is the implication that the psychoanalytic tradition endorses a *mixed* view of time. How is it possible to "mix" subjectivist and objectivist models? How can Freud, in particular, be considered to have a Kantian side? A subjectivist view of time seems to conflict with the popular understanding of Freud as a linear and continuous theorist.

Freud as a Mixed Model Theorist

As the first great personality theorist, Freud is crucial to our historical understanding of the field. Given the preceding treatment of Locke and Kant concerning linear time, many psychologists would undoubtedly place Freud's theorizing in the Lockean (and Newtonian) objectivist camp. Indeed, Freud is often viewed as the prototypical linear theorist. Even personality texts interpret Freud in this manner. Ryckman, for example, says that Freud held that "only past events determine the person's behavior."[32] Ewen explains that Freud "posits, as does Skinner, that all mental and physical behavior is determined by prior causes."[33] And Maddi, in true Lockean fashion, describes the formation of the ego and superego as determined by the accumulation of experience over time.[34]

It is no mystery why so many have construed Freud in this manner. He certainly appears to have many linear elements in his theorizing. On many occasions, for example, he cited his belief in *psychic determinism*: "Psycho-analysts are marked by a particularly strict belief in the determination of mental life. For them there is nothing trivial, nothing arbitrary or haphazard."[35] As Cameron and Rychlak correctly note, this makes Freud a "strict or hard determinist."[36] Even the seemingly inconsequential Freudian slip is assumed to be completely determined by the dynamics of the unconscious.

Freud was also known to have been quite sympathetic with Darwinian theory, both in the sense of animal urges and in the sense of life span "evolution." This meant to Freud that the first or earliest events in a person's life are the most fundamental. Oral experiences, for instance, dominate ego development because these are the earliest experiences of the infant. The importance of the Oedipus complex is also derived from its placement in time. Because it is thought to be one of the first social encounters of a small child, it is considered central to the eventual development of the superego. With this type of theorizing, it is not difficult to see the objectivist strains in Freud's explanations.

What about *subjectivist* strains? Is it possible that Freud incorporated some Kantian influences as well? This question can not only be answered affirmatively, but it is likely that Freud's theorizing was *dominated* by subjectivist and Kantian influences.[37] As Freud put it:

> Just as Kant warned us not to overlook the fact that our perceptions are subjectively conditioned and must not be regarded as identical with what is perceived though unknowable [that is, we never know the noumenal realm] so psychoanalysis warns us not to equate perceptions by means of consciousness with the unconscious mental processes which are their object. Like the physical, the psychical is not necessarily in reality what it appears to us to be.[38]

Freud is clear here that the conscious realm of our psyche is "subjectively conditioned" in a Kantian manner. He also seems to imply that all perception—whether of the physical or the psychical—is so conditioned.

Does this extend to the unconscious as well? Freud distinguished between the unconscious and conscious because the former is inborn and initially unaffected by its environment (and the vicissitudes of past experiences). It is clear, however, that the unconscious, too, functions in a subjectivist manner. Just as Kant's categories of understanding color our experience of the world from birth, so Freud's "primary process" colors our unconscious experience of the world from birth.[39] The environment is not simply "taken in" to be etched upon the tabula rasa of the unconscious. The environment is structured and selected in terms of the organization of the id. We have no awareness of this structure, i.e., the id remains in the unconscious. But then we have no awareness of the categories of understanding either. A lack of awareness does not prevent the id from functioning as a primitive (primary) mental processor and organizing (or categorizing) events in its own way—"the perception of unpleasure."[40]

Is time one of these mental organizations? Although Freud believed that the unconscious itself is "timeless" (as are Kant's categories), he did view the conscious as the purveyor of time. Time is not, in this sense, separable from the mind in the objectivist sense. Consider this passage from Freud's *Beyond the Pleasure Principle*:

> The Kantian proposition that time and space are necessary modes of thought may be submitted to discussion today in the light of certain knowledge reached through psycho-analysis. . . . Our abstract conception of time seems . . . to be derived wholly from the mode of functioning of the system *W-Bw* [perceptual consciousness].[41]

This seems to place Freud squarely in the Kantian subjectivist camp. Time does not occur apart from our consciousness; it is "derived wholly" from consciousness.

The problem is that Freud appears to rely upon objective time for his many intrapsychic structures. If the ego and superego, for instance, are considered to develop across time, how can Freud be a temporal subjectivist? Part of the answer lies in Freud's contention that these entities are formed *from* the id and are, in the final analysis, subservient to it. Freud once referred to the ego as the "organized portion of the id,"[42] and the superego, in turn, as a portion of the ego.[43] This means that the ego and superego are manifestations of the same entity. Time merely helps the id to adjust to society in meeting its needs; it does not change its basic identity over time: "The id is not concerned with the passage of time or of changes that may take place."[44]

Considering the id (and its ego and superego manifestations) as a subjectivist "category of understanding" implies that persons are structuring their experiences in an "id" manner—with "primary process" thinking—even when this process is extended to ego endeavors. As with Kant, environmental events do not come "in" to be stored with other environmental events from the past, at least not in the Lockean sense of a tabula rasa. Because all events are conceptualized by the person, they are *phenomena*. That is, they are not reality per se but the person's interpretation of reality. Early experiences can still be fundamental to the person's eventual personality (as Freud claimed), but these experiences are not "inputs" or etchings upon a tabula rasa, they are phenomena that the person has already subjectively organized.

Such theorizing often raises questions about the neonate. Modern Lockeans can usually accept that the mind structures the environment in some manner, but this mental structure has to have been put "in" by the environment sometime *following birth* to be consistent with a tabula rasa conception. This is one of the basic differences between the two philosophies. The Kantian has some means of mentally structuring the environment *from birth* and at every moment thereafter. Although Locke postulated innate mental qualities, it is clear that the simple sensations emanating from the environment hold primacy over inborn and eventual cognitive structures. Reflective operations are inborn but passive in relation to the "atoms" of sensation.

Lockean theorists are thus relegated to a linear style of explanation. They must presume (*à la* the tabula rasa) that environmental sensations are received sometime in the past, before a cognitive organization is available for mentally structuring the environment.[45] Freud, on the other hand, presumed that the person is an active, albeit primitive, conceptualizer in the first years of life. Primitive cognitive organizations, such as the pleasure principle, exist before environmental input, and are activated to some degree at each moment of perception. Cameron and Rychlak sum up this aspect of Freud in the following statement: "All [Freudian] theoretical speculations suggest an active, self-selecting human intellect, beginning in the first days of life and carrying forward from then on."[46] This places Freudian theory primarily on the side of a Kantian, subjectivist model.

One implication of this is that a Freudian personality is never "shaped" or "conditioned" by the environment, at least not in the Lockean sense. Again, several commentators have seemed to interpret Freud in this more Lockean light. Karl Menninger, for instance, has stated:

> Freud showed that men are extremely unequal in respect to endowment, discretion, equilibrium, self-control, aspiration, and intelligence—differences depending not only on inherited genes and brain-cell configurations but also on childhood *conditioning.* [47]

As Cameron and Rychlak observe, however, even the ego portion of the personality is "not as a 'shaped' outcome of environmental manipulation."[48] The ego is the socialized extension of an innate category of understanding. To be "socialized," of course, it must be influenced by its environment, but this influence occurs through phenomenal experience, not Lockean "atoms" of reality. From this Kantian perspective, the present state of the mind has as much to do with the *experienced* (and structured) "environment" (phenom-

ena) as it does the environment itself (noumena). The Lockean
"linearization" of Freud has misinterpreted his theorizing in its
own image. It has placed the environment in charge of Freud's
psyche—structuring and determining it—rather than the psyche
structuring and determining, at least to some degree, the environment.

This subjectivist side of Freud also sheds new light on his
notion of *psychic determinism*. Is the psyche determin*ed* or the
determin*er*? To a Lockean, the objectivity of time implies that determinism must happen across time. All causal and lawful processes must occur in temporal sequences. The psyche, in this sense,
is always determin*ed* by previous events—it cannot be otherwise.
Rychlak, however, has demonstrated that Freud employed many
other forms of determinism than this sequential variety. Besides
linear determinism, he used teleological and structural forms of
determinism in his theorizing.[49]

Briefly, structural forces are considered to operate simultaneously (rather than sequentially), having to do more with the
gestalt or context of concurrent events than with the linear sequence of events in time. Psychic events, then, are "determined" by
the part they play in the simultaneous psychic whole. Alternatively, telic forces "determine" psychic events through the purposes
and future goals of the psyche. The psyche is thought to "behave
for the sake of" its wishes and cathexes. Although this might seem
to require a linear future to have such goals, teleology actually
works through the present image or anticipation of a "future," and
thus requires no objective time.[50]

When other forms of determinism are recognized, interpretation of Freud's psychic determinism becomes more complex. Which
type of determinism is most fundamental to his use of this term?
Perhaps all types are, given his mixed model. It *is* clear, however,
that he considered the psyche to be more than merely another link
in the causal chain begun by the environment. In this more Lockean
sense, the psyche is little more than a way station for past experiences that are actually the determinants of the person's actions.
One of the advantages of nonlinear determinisms—structural and
teleological—is that they permit the psyche to truly *be* a determining factor. With structural determinism, for example, the organization of the mind forms a whole that *concurrently* determines its
parts. The whole can transcend the previous relations of its parts
and thus attain a new result. Freud, for example, sought "to understand [phenomena] as signs of an interplay of forces in the mind . . .
working *concurrently* or in *mutual opposition*."[51] This type of syn-

chronous causation means that the psyche is as much the determin*er* as it is the determin*ed*.[52]

Freud can also be seen to advocate a Kantian style of transcendence. Freud obviously felt that the phenomenal past is indispensable to an understanding of the present, but he also believed that people are not always determined by that past. People are not determined by their childhood traumas, fixations, and unresolved Oedipus complexes. If they were, personalities would be set in stone and psychoanalysis itself would be ineffective. Only a behavioral style of *directive* therapy would be possible with such Lockean theorizing—manipulating the environment to manipulate the person. But this was clearly not the style of therapy practiced by Freud. Not only do classical psychoanalysts *not* attempt to manipulate a person's environment, they are usually *nondirective* and interested in the *discontinuous* change of their clients' perceptions of the past and present—insight!

Such insightful transcendence is possible for Freud because one can reconstruct one's past. In a letter to his friend Wilhelm Fliess, he discussed this ability:

> As you know, I am working on the assumption that our psychical mechanism has come into being by a process of stratification: the material present in the form of memory-traces being subjected from time to time to a *re-arrangement* in accordance with fresh circumstances—to a *re-transcription*. Thus what is essentially new about my theory is the thesis that memory is present not once but several times over.[53]

Just as one constructs one's present in a Kantian (and Freudian) phenomenology, so one constructs one's past. In this sense, it is never the events of one's past that determine a person's personality but rather the *meaning* of one's past as (re)constructed in the present. This meaning is not itself determined by previous inputs because the meaning-giver (the id) existed before the inputs. This implies that the person has some measure of influence over the meaning of the past. The problem is that this meaning is not accessible, being in the realm of the unconscious. Freud's solution was not to manipulate the environment (the simple sensations from reality) as a Lockean would, but to bring the unconscious meanings into the conscious realm. This allows the control that we *already* exert over our unconscious meanings to become *conscious* control, and thereby permit other meanings to be assigned to the past.

In conclusion, then, Freud is truly a mixed model theorist. Portrayals of him as an exclusively linear or Lockean theorist have emphasized only one side of his complex theoretical scheme. Our discussion above has attempted to incorporate both models, though it has necessarily emphasized the Kantian elements in his theorizing, because these appear to be less understood. Three elements in particular have been misunderstood. First, although early experiences remain crucial, these are experiences of *phenomena* and thus are not the environment "etching" itself upon the person's mind. Second, other forms of determinism are evident in Freudian theorizing which are not tied to absolute time and which permit an escape from the *determinism* of the past in some instances.[54] One method of effecting this escape is, of course, therapy. However, Freudian analysis does not so much attempt to escape the past as to reconstrue it. This implies that the *meaning* of the past is the crucial issue, and a Kantian subjectivist view of linear time is a vital part of Freud's theorizing.

The Objectivist Tradition in Personality

As the first formal personality theorist, Freud set the precedent for theorists to follow. Theorizing was to be global and all-encompassing; theorizing was to have therapeutic implications, and so on. Many theorists attempted to follow Freud's lead (Jung, Adler), but when such theorizing was imported to America—where British empiricism and temporal objectivism prevailed—the response was less than enthusiastic. The subjectivist side of psychoanalysis, particularly, appeared loose and unscientific to many. (Note, of course, that scientific meant a Newtonian approach to explanation and method.) Still, psychoanalysis seemed to be catching on. Some psychologists wondered whether Freudian constructs could be redefined in more temporally objective terms. This would make them more palatable to psychological colleagues as well as more "scientific."

Dollard and Miller are perhaps most illustrative of those theorists who effectively "linearized" Freud. The word "effectively" here is meant to convey that most of Freud's constructs were successfully transformed into Lockean, objectivist constructs. Nevertheless, Dollard and Miller by no means captured the essence of Freud's theorizing. The temporal subjectivist side of Freud was completely lost, and though this was admittedly only one aspect of his theorizing, Freud's ingenious mix of the two temporal assumptions was lost as well. Let us examine some prototypical examples of this Lockean approach to personality. We will also attempt to contrast it with Freud's blend of the two sets of assumptions where possible.

Dollard and Miller retained both inborn and learned types of influences in their explanation of personality, but these influences are clearly rooted in a Lockean framework. They viewed the fundamental organization of the mind and behavior as stemming from environmental associations that are learned in the person's past. "Human behavior is learned; precisely that behavior which is widely felt to characterize man as a rational being, or as a member of a particular nation or social class, is learned rather than innate."[55] These learned associations are themselves imposed upon the person through stimulus and response connections.

Stimuli here are analogous to Lockean atoms of sensations, and responses are the later effects of these antecedent influences: "A response is any activity within the individual which can become functionally connected with an *antecedent* event through learning; a stimulus is any event to which a response can be so connected."[56] These stimulus-response associations lead eventually (across time) to habits that are then formed into a hierarchy of behavioral patterns. This response hierarchy is influenced by innate factors because those responses that reduce inborn drives assume the most favored position in the hierarchy. Nonetheless, the mind is still the result of environmental associations which are built up across time and controlled by the contiguity of objective time. The mind "stores" these associations in a classical Lockean preservation of past influences.

The contents of the mind and the patterns of behavior are always the *effects* of prior causes (i.e., responses). To be sure, these effects become causes themselves or the "cue-producing responses" of later effects. Language, for example, is the "human example *par excellence* of a cue-producing response."[57] This means that language, such as a label for a psychiatric symptom, can be the cause of some other effect, notably therapeutic benefits. However, the fact that language is itself a response reveals its linear origins, ultimately linking it with the past environment of the individual. Unlike Freud, then, this view of personality has no capacity for deviation from the linear past. Therapeutic benefits are not derived from changing the meaning of the past (as with Freud) but from finding ways of accessing other stored responses from the past.

With Dollard and Miller, all of Freud's theorizing takes on this exclusively linear quality. The *ego* is the store of past stimulus-response associations, and the *superego* is similarly deemed the accumulation of all socially learned behaviors. All these stored relationships—in this case stimulus/response associations—are formed to an important extent by the temporal relationships of the environment. Events that are supposedly contiguous in objective time (and space) are associated together and stored for later mediation by the ego and superego. Neither of these theoretical entities can organize reality *until* they themselves are organized *by* reality early in life (tabula rasa). Innate drives influence what relationships are most frequently used in behavior, but these are not organizing "categories of the mind" that form the relationships initially.[58]

Likewise, *insight* is not a discontinuous leap into a new meaning for the past, as it was with Freud. When the mind is exclusively an effect of past influences, it cannot turn on its causes and modify them subsequently. It can only proceed along its linear chain of events. Dollard and Miller accounted for the seeming discontinuity of insight by postulating that a new section of the response hierarchy had been accessed (usually via a label) to the present stimulus situation. This is akin to providing a computer with access to another part of its memory. Past programming is not changed, it is merely employed in a different manner. Therapeutic *interpretation*, then, is the provision of these labels, and *transference* is the positive or negative response style of the patient, given the stimulus of the therapist. In this manner, all of Freud's subjectivist elements were eliminated, and his objectivist elements brought to a commanding position.

Dollard and Miller's style of theorizing has been called "transitional," because it seemed to lead from Freud to more expressly objectivist conceptualizations.[59] Learning theorists wondered why they should retain the trappings of a Freudian conception at all. Why not simply drop the Freudian terminology and explain personality with strictly objectivist principles? B. F. Skinner is perhaps best known among the theorists who did precisely this. He used learning principles exclusively to explain how persons are shaped into their current personalities (behavior), and he applied these principles in programmed instruction and behavior therapy. Both of these applications involve the gradual (across-time) modification of previously conditioned responses so that the number of responses leading to reinforcement is increased.

Newtonian and Lockean assumptions permeate Skinner's conceptions. Conditioning principles alone—whether operant or classical—require temporal contiguity between the behavior/reward and conditioned/unconditioned stimulus, respectively. The mind does not create this contiguity according to Skinner. Linear time and the events that occur within it are objectively dictating these contiguous relationships. The notion of shaping, too, requires events along the line of time. A person's behavior is molded by repeated exposure to the various reinforcements of the environment, all of which require linear sequence and accumulation of the past:

> In a behavioral analysis, a person is an organism . . . which has acquired a *repertoire of behavior*. . . . [A person] is not an originating agent; he is a locus, a point at which many genetic and environmental conditions come together in a joint effect. As such, he remains unquestionably unique. No one else (unless he has an identical twin) has his genetic endowment, and without exception no one else has his *personal history*. Hence, no one else will behave in precisely the same way.[60]

In a remarkable exchange with Carl Rogers, Skinner revealed how willing he was to apply these explanations to his own behavior. As Rogers reports, he questioned whether Skinner

> actually made certain marks on paper and emitted certain sounds here simply because his genetic makeup and *his past environment* had operantly conditioned his behavior in such a way that it was rewarding to make these sounds, and that he as a person doesn't enter into this. In fact if I get his thinking correctly, from his strictly scientific point of view, he, as a person, doesn't exist. In his reply Dr. Skinner . . . stated, "I do accept your characterization of my own presence here."[61]

Even Skinner's prodigious achievements in therapy and education carry this Newtonian stamp. The learning and shaping of new behaviors (through conditioning) take place across time. Therapeutic and educative change occurs gradually through "successive approximation"—no discontinuous jumps from one learning stage to another is allowed. Reductionism is then possible. Because the process of change takes place smoothly across time, it can be reduced to the successive steps necessary for learning to take place. This means that anybody can learn even complex behaviors. All the teacher or therapist needs to do is break the complexity down to its component parts and teach them one at a time, eventually chaining them together for the completed product.

Skinner was also a foremost advocate of *universality*. He proposed that all behaviors could be understood in terms of general laws of science which are independent of specific context. Indeed, most of his professional life can be characterized as the pursuit of these laws and principles. Whether it be successive approximation, schedules of reinforcement, or the principles of operant conditioning, Skinner argued that such processes are the same, regardless of their temporal context. The content of the various laws of learning might themselves vary, but the processes behind these variations are constant and thus universal. In fact, Skinner held that such processes are not only continuous across the person's life span but also continuous across the species that make up the person's phylogenetic evolution (i.e., phylogenetic continuity).[62]

Bandura's Social Cognitive Theory

Recently, objectivist theorizing has seemed to adopt a more conciliatory approach to subjectivist concerns. In some cases, theorists have appeared to satisfy these concerns, yet remain fundamentally Lockean and thus "scientific" in the Newtonian tradition. Perhaps the theorist most noted for this is Albert Bandura. Bandura has gone so far as to advocate human agency and freedom—capabilities that have long been thought to be impossible in the objectivist tradition. Objective time and causality have been thought to obviate freedom of will.[63] Bandura, however, criticizes those personality theorists that propose "one-sided determinisms"—theoretical analyses that result in personal factors being determined by environmental factors. He finds that "in the regress of prior causes, for every chicken discovered by a unidirectional environmentalist, a social cognitive theorist [Bandura] can identify a prior egg."[64]

Bandura here is bucking one of the central constructs of the behavioral tradition—conditioning or reinforcement history. As Bandura explains, strict behavioral conceptions, such as Skinner's, have been increasingly brought under critical fire. Instead of "antecedent stimuli" accounting for all behavior, "personal determinants" have been revealed to play a larger and larger role.[65] Behaviorists have been quick to counter with other Newtonian temporal regressions. Many grant the existence of so-called personal determinants (e.g., cognitive behaviorists) but contend that these too are the effects of past experiences, just farther back in time. In this sense, the behavioral model of causation has been enlarged to include temporally distant as well as temporally proximal stimuli. According to Bandura, however, the "one-sided" determinism of the environment remains unchanged.

What Bandura advocates instead is his unique *triadic reciprocal determinism*. This form of determinism postulates that behaviors, personal factors, and environmental influences all operate interactively to determine one another (see Figure 3.1). He is quick to note that the term "determinism" is not meant to connote a prior sequence of environmental causes functioning independently of the individual. Even environmental factors are themselves affected by the self and behavioral factors. Unlike a Skinnerian conception, observes Bandura, people are not considered mere reactors to the environment. "People create environments and set them in motion as well as rebut them. People are foreactive, not simply counteractive."[66] Self factors, however, have no independence or autonomy either. Bandura criticizes existentialists here "who contend that people determine what they become by their own free choices."[67] True reciprocal determinism means that all three factors are involved in determining each other as they influence the person.

So far, Bandura appears more than just conciliatory toward the subjectivist view of linear time and human nature. He reveals a resistance to causal regression and a form of interactionism that, on the face of it, seems quite compatible with Kant. Before we can make this judgment, though, crucial questions need to be answered about the nature of this interaction. Skinner, for instance, proposes an interaction between environmental control and personal countercontrol. However, as Bandura correctly notes, this does not mean that Skinner has left the Lockean tradition—the environment is still ultimately on the causal side, and the person is still ultimately on the effect side of this "interaction." To move into the subjectivist tradition, a theorist must postulate some means of the person transcending (or escaping) the *deterministic* influences of the past. Kant had two such means: (1) an innate capacity to structure environmental information, at birth and each moment thereafter, and (2) an innate ability to formulate novel ideas cognitively, and thus transcend chains of causation (and past simple sensations) across time. Does Bandura's theorizing contain such conceptions?

It appears that the answer is no, though Bandura is not sufficiently clear on the issue to make this determination unequivocal. Let us explore first the issue of causal chains across time. If a theorist asserts such linear chains, then it is likely that he or she is still a temporal objectivist. Linear chains are not possible without

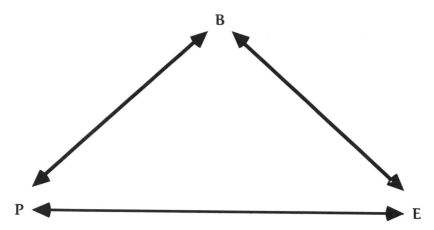

Figure 3.1   Bandura's schema of the relations among three
classes of determinants

an objective time frame in which to discern what is before and
after. Bandura, however, discusses *constructive* cognitive factors
which have been part of the structuralist tradition.[68] Could he be
advocating some type of structuralist interaction among the triadic
factors that involves their simultaneous relation—as a gestalt—
and thus transcends their past relations?[69] Bandura addresses this
issue directly:

> The triadic factors do not operate in the manner of a simulta-
> neous wholistic interaction. Reciprocality does not mean si-
> multaneity of influence . . . . It takes time for a causal factor
> to exert its influence. Interacting factors work their mutual
> effects sequentially over variable time courses. Even when
> bidirectionality of influence is almost immediate, as in verbal
> interchanges in which a question by one person evokes a
> prompt reply from another, an influence is not altered before
> it has exerted itself."[70]

This quote makes it highly probable that Bandura has *not* left
the objectivist tradition. Although three separate factors are involved in the interaction, each factor takes its turn in linear time,
and each determines the other in the usual cause and effect sequencing of Newton. Figure 3.1 shows Bandura's usual sche-
matization of the relations among the three classes of determinants. However, this schema can be misleading to a theorist who is

accustomed to linear time flowing in the figure from left to right. To insert the time dimension properly, as conceptualized by Bandura, one would have to pick an arbitrary starting point, and depict the three factors in linear manner of Figure 3.2. This figure makes it more apparent that standard linear causation is in effect. Although it is true that each set of determinants affects the others, each is also wholly determined by the previous effect. This makes deviation from the past—whether it be an environmental, personal, or behavioral factor—impossible because the past is clearly the determinant of the present.

Why then is Bandura so critical of those theorists who resort to causal regress or reinforcement history? The answer is that Bandura is not so much interested in linear time (or linear causation) as he is interested in the *non*interactional influence of the environment. That is, he is *not* concerned about causal regression

...E ⟶ P ⟶ B ⟶ E...

Figure 3.2   Bandura's schema across linear time

per se; indeed, linear causation is an objective fact for Bandura. He is instead concerned about one of the three factors escaping his brand of interactionism *through* causal regression. He knows that if the environment (or any other factor) is viewed as "first" in line, then he cannot avoid the implication—given his *own* linear view of time and causality—that this factor determines the factors which follow.

An important question arises in this regard: "Which of the three sets of determinants *does* comes first?" Although all determinants are themselves "determined" in Bandura's conception, the ultimate regress of causes is not a trivial question. Certainly, as Bandura repeatedly advises in his many writings, "for every environmental chicken, a social cognitive theorist can locate a personal egg." Nevertheless, this type of regression has to stop at some point, presumably at the moment in which a baby or fetus first begins to sense the environment. With Kant (and Freud), clear statements about inborn factors were expressed to address this issue. This made personal and environmental factors *simultaneous*,[71] and allowed for a subjectivist style of theorizing—always and at each discrete moment thereafter permitting personal factors to structure as well as be structured by the environment.

Bandura, however, says nothing about innate structuring abilities. He appears to fear that such abilities would place him in another sort of "one-sided" determinism—this time on the side of autonomous personal factors. If anything, he seems to lean toward the environmental side. This might seem at variance with his avowed distaste for environmental "chickens," yet as a linear theorist, he must hold that the organism *begins* at some point in time. Given his temporal sequencing of triadic factors, the organism can begin with only *one* of these factors (the reductive quality of any Newtonian framework). Because he does not explicitly postulate innate personal factors, statements like the following give a definite linear edge to the environment: "Having earlier *external* origins, of course, in no way detracts from the fact that, once established, self-influences operate as current contributory causes of behavior."[72]

Bandura's discussions of illustrative personal influences also carry this implication. Consider his personal factor of "self-criticism." First, Bandura does not postulate that self-criticism is in any way innate. Indeed, he makes clear that young children lack the personal standards necessary for this capacity. Expressions of self-criticism arise following parental discipline. Children learn standards from their parents because they are reinforced through a lessening of parental displeasure and/or the reinstatement of parental approval. This learning would undoubtedly include personal and behavioral factors as the linear chain of triadic processes unfolds across time. However, it seems clear that the development of self-criticism qua self-criticism is *begun* by and ultimately depends upon reinforcements and information emanating from the environment.

In a similar vein, Rychlak has analyzed Bandura's model of cognition as being *mediational*.[73] That is, cognition is a mediator between the input of the environment and the output of behavior, with the entire process operating across time and begun by the environment. This would mean, as Rychlak puts it, "that [Bandura] views mediators as originating more on the effect than the cause side of a cause-effect sequential interaction."[74] Certainly, we find many of Bandura's statements reflecting this emphasis: "Cognitive events . . . do not function as autonomous causes of behavior. Their nature, their emotion-arousing properties, and their occurrence are under stimulus and reinforcement control."[75] Such mediational theorizing may seem in contradiction to Bandura's claim that humans have transcendent capacities. He discusses at some length the capacity of people to be "free" and "generate novel ideas that transcend their experiences."[76]

Bandura's actual definitions of these capacities, however, are compatible with linear mediation. Freedom, for example, is "the exercise of self-influence"[77] which allows the environment to determine the self as a factor (e.g., the self as a storage system). It is also unclear what past experiences Bandura feels can be transcended. He consistently describes experience as "stimuli" quite analogous to Locke's "atoms" of sensation and depicts these stimuli in a causal sequence followed by higher mental functions and behavioral responses analogous to Locke. A Kantian phenomenology of experience, on the other hand, does not yield stimulus (or causal) power to any aspect of the environment alone. Stimulating aspects of our experience are stimulating, at least in part, because our mind has selectively attended to or endowed them with meaning. In this sense, a Kantian phenomenal interaction requires the simultaneous conjoining of cognitive and environmental structures, with neither having causal primacy over the other.

The upshot is that Bandura evidences neither of the Kantian requirements for a subjectivist theory. He postulates neither innate factors of cognitive structuring nor factors that permit true transcendence of the past. Indeed, he explicitly postulates a personality process that requires temporal sequence and duration. Even his conception of interaction has this property. Although the nature of the personal determinants which Bandura says "transcend" experience is not entirely clear,[78] it *is* clear that these determinants are a part of a causal chain stemming from the past. In this sense, personal determinants do not "transcend," at least not in the sense of escaping the determinism of the past. Bandura's transcendence has more to do with independent sources of interactional input, rather than a truly *original* contribution to the interaction itself.

The Subjectivist Tradition—Newtonian Anomalies

The explanatory power of the objectivist tradition should be readily apparent at this juncture. It subsumes all sorts of psychological phenomena, particularly those viewed in the empiricist tradition. In fact, some psychologists may wonder how one can escape temporal objectivism in personality theorizing. Perhaps it was Bandura's *intention* to escape from objectivism—with his emphasis upon transcendent capacities such as freedom—but his Newtonian metaphysics would not quite permit this. Freud, too, evidenced many strong strains of objectivism. Despite his Kantian philosophical heritage (as an Austrian) and predominant subjectivist style of thought, even he succumbed at points to the linear emphasis of Locke and

Darwin. Is there anyone in the personality tradition who *completely* escapes the Newtonian tradition?

Hall and Lindzey list several theorists whom they feel emphasize "contemporaneity" as opposed to "early experiences" in personality.[79] They include Allport, Rogers, Binswanger, Boss, and Lewin as thinkers who have resisted linear and continuous explanations in the favor of more present-focused and discontinuous descriptions. If this is true, these theorists would not fit the definitions of personality offered at the outset of this chapter by recent commentators.[80] Recall that these commentators required personality to be stable over time and the result of early experiences with the family.[81] Theories without these characteristics must then be judged as inadequate or unscientific. If, on the other hand, we recognize that such judgments stem from Newtonian criteria, then we can see these criteria as emanating from a *point of view*, and not rule out other theoretical options. Space limitations prevent our discussing all these options here. Instead, we will discuss the most straightforward theorist on the issue of time—Kurt Lewin—and explore the thinking of another theorist overlooked by Hall and Lindzey in this regard—Carl Jung.

The ideas of Lewin and Jung can truly be said to be anomalies to the Newtonian paradigm in personality. It is no coincidence that both theorists derived much of their ideas from outside the discipline of psychology, notably physics.[82] Most physicists have long ago abandoned their Newtonian framework for time and explanation, but few psychologists seem to be aware of this (see Chapter 1). Jung and Lewin, however, were very conscious of recent theoretical advances in physics. They recognized that the physical world (and its metaphysical assumptions) could not be ignored in psychological theorizing. They constructed comprehensive theories that were undergirded with *non*-Newtonian assumptions of time and explanation.

Kurt Lewin

Kurt Lewin believed the advances of the physical sciences cast considerable doubt on the notion that a causal chain from the past could produce events in the present. He distinguished between two types of causality: historic and systematic. Historic causation is essentially Newtonian in that it "gives an account of the course and interweavings of causal chains of events."[83] Systematic causation, on the other hand, is ahistoric. Its determinism of events "consists in the properties of the momentary life space or of certain integral parts of it."[84] Although he recognized the legitimacy of the historical, he felt that the systematic was the more fundamental for psychology: "This influence of the previous history is to be thought of as indirect in dynamic psychology."[85]

To support and amplify the systematic, Lewin formulated two principles: The *principle of concreteness* and the *principle of contemporaneity*. The first principle is essentially the proposition that abstractions cannot cause effects—only concrete factors can produce effects. This proposition may seem obvious, but as Lewin notes, its full implications for explanation are rarely acknowledged. Lewin defined "concrete" as "something that has the position of an individual fact which exists at a certain moment."[86] In other words, concrete factors are those that are literally "there," existing and present. In this sense, past and future states of an object are not concrete because they are not there, existing and present. As Lewin put it:

> We shall strongly defend the thesis that neither past nor future psychological facts but only the present situation can influence present events. This thesis is a direct consequence of the principle that only what exists concretely can have effects. Since neither the past nor the future exists at the present moment it cannot have effects at the present. In representing the life space therefore we take into account only what is contemporary.[87]

Lewin is here attempting to take seriously something that most of us take for granted—i.e., all we really and concretely have is the present. Although we can *represent* past events through our memories and recording devices, past events are not concretely existent at any (present) moment in time. The future is likewise not literally present, except in our anticipation or expectation. Therefore, neither the past nor the future (in a Newtonian sense) can cause present effects because only concrete factors can produce concrete effects. Part of Lewin's point is that our cognitive abilities are integral to our usual notions of causation. Without a memory, for example, we would probably never look to a prior "cause" for a present effect. The prior "cause" does not concretely exist in the present, so its image would not be preserved (in memory) to allow us to connect it with the "effect" now occurring. Newtonian causation, therefore, violates the principle of concreteness, because it contends that abstract events of the past—which can only be *repre-*

sented in the present—are causally linked with concrete events of the present.

Such theorizing may be difficult to comprehend at first, especially given the familiar Newtonian perspective on this issue. Many people in Western culture have endowed the past and future states of an object with an almost "concrete" status. Indeed, this helps account for a peculiarity in traditional behaviorist explanations. Although behaviorists have historically restricted their attention only to "observables," many behaviorists have given the *un*observable factor of "reinforcement history" considerable prominence in their explanations.[88] It is true, of course, that reinforcements *could* have been observed across time, but the "concrete" fact is—as Lewin would say—that this history is not *being* observed at the time in which the behavior occurs. The past holds such a status of realism in our culture that even the rigor of traditional behaviorism (and positivism) does not note that "reinforcement history" is not itself observable, or in Lewin's terms, concrete.[89]

Lewin's *principle of contemporaneity* makes his present focus even more lucid.[90] This is the notion that only the holistic interaction of concrete factors which are *contemporaneous* with the effect can produce the effect. Lewin's gestalt roots told him that the interacting gestalt of the surrounding context takes place simultaneously with the event itself. The ground occurs simultaneously with the figure, and in a real sense, produces or causes the meaning of the figure. When placed together with the principle of concreteness, this means that past and future events cannot cause events of the present. Only events that are contemporaneous can be causal, and past and future events only exist contemporaneously as abstractions in the minds of the observers (through memories and anticipations).

The principle of contemporaneity also cannot abide by the usual notion of linear time as a continuous flow. Lewin contended that the course of events must be represented by a series of "momentary sections" or slices of time that are discrete and separate.[91] This means that each moment is a gestalt—with all the "causal" factors concretely existent within that moment. The next moment is a reorganization of the first moment, and a potentially different gestalt. Sometimes, Lewin noted, the reorganization of the event and its context is so slight that the discontinuity of change is hardly noticeable, appearing as continuity. Nevertheless, the discontinuity or discreteness of the two moments is the basic reality, not its continuity, since at any moment discontinuous change is possible.

Lewin went on to develop these basic assumptions in his theory of personality. He observed that most personality theorists rely on the past or future, and see persons as continuous entities across time. Lewin also rejected the usual notion of the world "coming at" the person across time and space. Examples of this include the modern Lockean notion that the environment is "stimuli" or "input" moving from "out there" (in the environment) to "in here" (in the mind). Lewin felt that the relation between environment and mind is contemporaneous and that the two entities form a patterned whole which interact synchronously. Determinants of personality do not stem from the past in the sense of cognitive storage, nor even the immediate past with the person as a "responder" to stimuli or input. Personality is the gestalt of the person and environment in the present, each giving meaning to the other as parts give meaning to the whole. This is the reason that Lewin stressed group, as opposed to individual, therapy—the group could form a contemporaneous whole (see Chapter 7).

Carl Jung

Carl Jung offers an alternative means of escaping Newtonian assumptions in personality theory. Jung might seem a surprising inclusion here given his psychoanalytic heritage. Psychoanalysis has long meant historical causation to most psychologists,[92] and Jung is often considered a mixture of objectivist and subjectivist themes.[93] Nonetheless, when Freud's subjectivist theoretical strains are understood (see above), Jung's intellectual heritage and subjectivist emphasis are not so surprising. Jung was even more dissatisfied than Freud with objectivist views of time, particularly later in his career. He felt that most scientists ignore data which are incompatible with an objectivist framework. If events do not accord with linear causal laws, they are given pejorative labels like "error variance," "coincidence," or "chance." Jung noticed, however, that many of these events are meaningful and, indeed, make up the vast majority of human observations. To make sense out of these, he postulated a principle of explanation quite unlike that of absolute time: *synchronicity*.

Jung provided definitions of synchronicity at two levels of abstraction. At a more concrete level, synchronicity is simply a meaningful cross-connection, a relation between events that is significant to the observer but inexplicable from a Newtonian framework.[94] Jung's more abstract definition of synchronicity was the "psychically conditioned relativity of space and time."[95] In other words, his view of time was strictly Kantian. He saw space and time as intellectual constructions, psychic in origin, and not existent in themselves. Space and time are "only apparently properties of bodies in motion and are created by the intellectual needs of the observer."[96] This actually parallels Einstein's views to some extent (see Chapter 1). Einstein also considered time and space to be "free creations of the human intelligence, tools of thought, which are to serve the purpose of bringing experiences into relation with each other."[97]

One of the consequences of postulating the subjectivity of linear time is that meaningful connections do not require temporal sequence or duration. Causal relationships can occur simultaneously (or "contemporaneously," to use Lewin's term). Jung noted that simultaneous relationships are impossible to explain from a linear causal account, because there is no time for the energy of transmission between events. In other words, causally related events from a linear perspective *must* be sequential to allow for some type of energy transmission to travel between each event. Truly simultaneous events—especially if separated in space—cannot be related by such an energy, and thus cannot be meaningfully explained from a linear perspective. Jung, however, was convinced of the importance of simultaneous relations and coined the term "synchronistic" to describe them.

Of particular interest to him in this regard were the parapsychological experiments of J. B. Rhine and Jung's own experiences in therapy. As an example of the latter, he cited his own patient who described a dream involving a scarab.[98] At the time in which the patient was describing the scarab—the most crucial element of the dream—an odd tapping occurred at the window of the therapy room. Jung turned around to see a flying insect knocking against the window from the outside. He opened the window and caught the creature—a scarab beetle. The scarab had apparently felt an urge to enter a mostly dark room—in contrast to its usual habits—at the same moment as the patient's description of the scarab in the dream.

Jung was quite aware that most people would dismiss this as mere coincidence. Nevertheless, Jung asks us to consider whether this is *really* "mere coincidence" or the Western thinker's need for a linear connection between the two events. That is, in order to be "scientific"—at least in the Newtonian sense—some linear explanation is required to connect the beetle with the description of the dream. If these two events are, in fact, related to one another, a linear explanation would have to deny the simultaneity of the events because one event would need to be the cause and the other event the effect (and thus sequential in time). A linear explanation would also require some sort of force or energy to be transmitted between the two events because all the causal time and space would need to be "filled" to have the effect. Without such an explanation, most Western thinkers would reject out of hand any meaningful relation between the two events.

The problem with this rejection is that many findings in modern physics, particularly quantum mechanics, have the same nonlinear nature as the example Jung cited. Consider the fact that two previously correlated particles can be split apart, travel some considerable distance from one another, and yet remain simultaneously related.[99] As soon as one particle is disturbed by an observation, for example, the second particle immediately assumes a similar value of disturbance. The two objects can be light years apart, but as soon as one is affected, the other is synchronously affected. Because of the simultaneity of this relationship, physicists have had immense difficulty connecting these events with linear forces traveling from one to the other.[100] Still, this is often the only explanation Western thinkers have to offer,[101] which is precisely Jung's point.

Jung also felt that we are emotionally vulnerable to synchronistic relations, whether or not we can explain them. The beetle in Jung's window was psychologically significant to him and his patient, regardless of their ability to explain the connection between the events. Jung held that such synchronistic relationships affect us everyday. If we insist on ignoring them as "chance" or "coincidence," then we continue to be unaware of their influences. If, on the other hand, we recognize their importance, we are better able to understand their effects. Indeed, Jung asserted that our own psychic and emotional systems are themselves constructed with synchronistic relationships. Synchronicity is, in fact, foundational to his view of the unconscious, complexes, and archetypes.

Conclusion

The conceptual diversity of personality theory has been shown to include assumptions of time. Theorists range widely on temporal issues, from objectivism to subjectivism. Some theorists, such as Dollard and Miller, Skinner, and Bandura follow the objectivist style of John Locke. Locke not only affirmed the teachings of his friend Newton, but his philosophy seems to have embodied many of the epistemological implications of Newton's framework for explanation. On the other hand, some personality theorists, such as Lewin and Jung, follow the more subjectivist style of Kantian philosophy. Emphasis is placed on the present for explanation. The past is by no means ignored, but it is less likely to be viewed as the sole determinant of the now.

Freud is perhaps most properly seen as a mixed model theorist. There is no mistaking his stress upon the past when making sense out of the present. However, this has led many commentators to presume mistakenly that Freud was a temporal objectivist exclusively. To be sure, Freud valued past-to-present explanations, but he did not hold that the past invariably determined the present. Strong strands of temporal subjectivism are visible in his theorizing. Indeed, without these strands, Freud's particular brand of therapy would be impossible. His therapy was intended to help patients reconstrue the meanings of the past through their insight into unconscious meanings rather than overcome the past through a behavioral reprogramming of old inputs. Therefore, an overlooked aspect of Freud's theorizing is the transcendent capacities that made his brand of therapeutic change possible.

Bandura, on the other hand, has highlighted transcendent capacities in his own theorizing (e.g., freedom). An analysis of his temporal assumptions, though, shows that it is unlikely that transcendence (in the sense of escaping causal chains from the past) is possible. Bandura discusses how each portion of his triadic reciprocal determinism makes "novel" contributions to their interaction. Yet each portion is apparently an effect of a previous cause—which itself stems from another portion of the reciprocity. How cognition, for example, makes a *novel* contribution when it is wholly caused (in the previous instant) by environmental or behavioral stimuli is unclear. In any case, Bandura's description of the temporal relations among the constituents of this interaction, along with his continued emphasis upon stimulus-response chains, make it evident that he affirms an objectivist view of personality.

In contrast, Lewin and Jung offer an essentially subjectivist portrayal of personality. Chains of causation across time are deemphasized, if not entirely avoided, as the main ingredients of personality. Interactions are not sequential, as they are with Bandura. They are simultaneous. This means that no part of the interaction (e.g., environment or cognition) has causal (or temporal) primacy. *All* parts are necessary to the meaning of the whole.

This also means that personality is "determined," but not in the linear sense. Personality is more the product of the present; the meaningful relations between the parts "cause" the meaning of each part and thus personality itself. This meaningful relation can, of course, change in the next moment, and this implies that one's personality can change discontinuously without any entity or force-across-time being responsible.

The variety of temporal positions taken by personality theorists makes it imperative that issues related to a Newtonian framework be brought to the forefront. As mentioned early in the chapter, few personality commentators acknowledge the diversity of temporal explanations, even in their own field.[102] Many recent texts, in fact, implicitly advocate a Newtonian perspective on personality theorizing. Their frameworks for personality make the theorizing of Lewin and Jung seem inexplicable, if not fantastic. Indeed, most recent texts have dropped any description of Lewin altogether, and Jung's explanations are often characterized as "mysterious processes."[103] If these Newtonian frameworks are allowed to dominate the field, conceptual misunderstandings will persist and theoretical options will be overlooked. Subjectivists like Lewin and Jung will continue to be viewed as unscientific, when the only thing that is unscientific about them is their lack of a Newtonian conceptual pedigree.

Chapter 4

PSYCHOLOGICAL METHOD

Time not only permeates psychological theory, it also permeates psychological method. As methodologists, McGrath and Kelly, observe, "temporal factors are ubiquitous in our methodology."[1] Measures of reaction time, stimulus duration, and rate of change all depend upon assumptions of time. Psychology's research designs are profoundly temporal in character. Researchers must decide upon the length of their experiments, the chronological order of experimental tasks, and the interval between treatment and observation. Even so-called threats to experimental validity—e.g., history, maturation, and regression[2]—require assumptions of time.

With the need of such assumptions, it should not be surprising that researchers have adopted a Newtonian perspective on time. Recall from Chapter 1 that psychologists originally looked to Newtonian method as the proper way to conduct science. Newton's own method, of course, was infused with linear time. He presumed that linear time existed as a separate medium to measure the motions of the world,[3] so he specifically formulated his method to exploit this temporal nature of reality.[4] Observations across linear time were gathered to discover the chronological order of natural events. Those events that occurred "before" were candidates for causes,[5] and those linear sequences that occurred repeatedly (across time) were candidates for laws.

Mainstream psychological researchers have adopted virtually every aspect of this approach to method. The purpose of this chapter is to explicate this temporal paradigm as well as note any methods that are anomalous to the Newtonian framework. The chapter begins with a brief history of method as it relates to linear time and provides a needed background to the issues at play. The chapter then demonstrates how psychology's methods depend upon each of the five characteristics of a Newtonian framework, as delineated in Chapter 1. Lastly, several anomalies or exceptions to this paradigm are described. These take two major forms: empirical phenomena that do not seem amenable to linear methods and whole methods that have successfully moved away from a Newtonian framework.

The History of Method and Time

Perhaps surprising to today's methodologist, the earliest methods for investigating the world were formulated without the need of conventional time assumptions. Method required neither temporal order nor temporal duration in the Newtonian sense, and thus never manifested the characteristics of Newtonian explanation so endemic to psychology today. The first methods were based upon simple human reason or logic—"thinking it through." If we define method as "a procedure for determining whether a proposition is true," logic accomplished this purpose by testing propositions through implication and conclusion. Aristotle was one of the first to formalize this method. His syllogistic mode of reasoning is an excellent illustration of how such a method escapes conventional linear time.

Consider the classical syllogism: All men are mortal. Socrates is a man. Therefore, Socrates is mortal. Although articulating this syllogism seems to require time, the relation between the two premises—the conclusion—does not. That is, simultaneous with an understanding of the first two sentences (the premises) is an understanding of the third sentence (the conclusion). A logical process has been conducted—from premise to conclusion—without linear time having passed. Figure 4.1 illustrates this by depicting the two premises as circles within circles. When the premises are drawn in this fashion, the logical conclusion is immediately evident (i.e., the innermost circle is within the outermost circle). No interval of objective time is necessary to explain the operation of this method.[6] Indeed, linear order is also unnecessary. The order in which the circles are drawn has nothing whatsoever to do with the conclusion reached. The second premise could be drawn first and not affect the conclusion. Thus, their temporal order indicates no linear primacy.

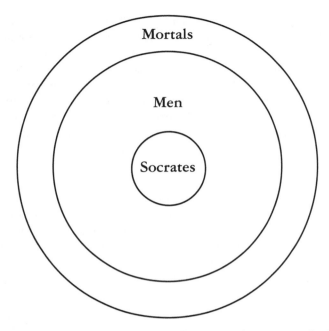

Figure 4.1.   A pictorialization of the simultaneous relations of a logical syllogism

Logic was formalized even further when geometry was developed. As Bertrand Russell has demonstrated, logic and mathematics are virtually identical enterprises.[7] Mathematics is, from this perspective, a precise articulation of logical relations (equal to, greater than, etc.), and the classical syllogism of Aristotle can itself be expressed in mathematical terms. This, of course, means that mathematics is also devoid of linear time. For instance, a change in the problem on one side of the equal sign simultaneously affects the solution on the other side (without objective time having passed). In this sense, mathematical and geometrical relations are tautological and thus implicit in the same manner that Aristotle's conclusion was implicit in his premises.[8]

Geometry was a monumental advance in method. Although geometry was, in many ways, an extension of traditional human logic, its greater precision and quantifiability permitted easier application to the outside world. Indeed, geometry became so successful as a method that its connection with human reasoning was soon forgotten. Geometry became reified as an entity on its own in the external world. In other words, geometry did not just test propositions about the world, the principles behind the regularity of the world were *themselves* geometrical. This began the objectivity and universality of method. Geometric patterns of the universe were not affected by the passage of linear time or the particular conditions under which they were observed or discussed. They were thought to be "universal," cutting across all times and spaces.

Universal methods were also supported by the theologies and philosophies of their day. Theologians saw the world as a manifestation of God's rationality, and the epitome of such rationality was geometry. Similarly, philosophers such as Plato believed that eternally perfect forms were behind the changing world. The world might seem to be in a state of flux, but this was a superficial analysis according to Plato. "Behind" this flux was the universal world of unchanging geometric forms and principles. Planets, for example, were considered to orbit in perfect circles throughout time, sometimes in spite of information to the contrary (e.g., Kepler's). The point is that all the major disciplines at this time—science, mathematics, philosophy, and religion—converged upon a universalist view of truth.

Another method, however, began to take shape as early as the thirteenth century. Purely logical and mathematical methods were increasingly losing their luster. They were seen to rely too heavily upon verbal argumentation and rational self-evidence (e.g., logical proofs or geometric theorems). Roger Bacon and later, William of Ockham, argued that knowledge could be gained through direct experience of the world, rather than contemplation or mathematical derivation.[9] Yes, the world of flux changed constantly. But did not the flux itself deserve greater attention? Logical and mathematical methods were too abstract and distant from the world. It was high time that knowledge of the world was itself connected to actual and systematic observation of that world.

Two methods were thus vying for favor at this time: rationalism and empiricism. Rationalists tested propositions through informal and formal reason. Relations—whether logical or mathematical—were already implied, and constituted timeless context-free "laws," much like reason and mathematics themselves. The empiricist, on the other hand, approached method with an appreciation for the "real" world. The truth about the world was not already implied or self-evident as it was for the rationalists. Truth was derived from the data of the world itself. Even if the world *were* in constant flux, argued the empiricists, this did not negate the need for a full understanding of that flux.[10]

Newton's brand of science arose from a unique wedding of these two methods. The philosopher Descartes is perhaps most responsible for the matchmaking that led to this wedding. Descartes essentially put the timeless universals of the rationalists "in motion." He took the invariant theorems of Euclidian geometry and gave them a dynamic character.[11] A circle, for example, was defined before Descartes as a static line whose points were equidistant from a given center. No motion (or change across linear time) was necessary to understand or employ this geometric shape. Descartes, however, redefined the circle as the *motion* of a point moving a constant distance from the center. Although he still presumed the circle to be universal and "behind" phenomena of the world, his definition endowed this universal with a definite temporal quality. Instead of static and unmoving principles as truth, truth was associated with regularities and patterns of motions across time.

This notion was so appealing that most thinkers of the sixteenth and seventeenth centuries assumed it implicitly. Galileo and Newton considered such temporal universals to be the foundation of scientific knowledge.[12] When Newton studied the universe, he focused upon its linear movements almost exclusively. His three laws of motion (i.e., displacement across linear time) stand as testament to his abiding faith that motion is the cornerstone of any knowledge base in science. Finding the universal truths in such motions had been mightily simplified by Descartes. One merely subjected the pattern of movements to Cartesian geometry. This was a crucial step in our brief history of method because it was now possible to keep the universality and rationality of truth, and yet apply it to the world of change and motion.

What is rarely recognized in this step, however, is the role of linear time. Before this step, geometric universals required only space to conceptualize them. Space was viewed as having no content, and thus not affecting the entities within it except in uniform and consistent ways. Space differed only in quantity and not quality. The addition of motion, however, required a before and after and some quantifiable means of measuring the interval between these two points. Motion, therefore, required an additional frame of reference much like space. Motion required a before and after frame of reference that flowed "without relation to anything external,"[13] as Newton put it, because the motion of external objects was to be measured by it. *Absolute time* was thus born and, like absolute space, was uniform, without content, and the bearer of events within it.

The "mechanization of the world picture," as some historians have phrased it,[14] provided further impetus to this temporal method. As conceptualized by Galileo and Descartes, machines (e.g., clocks) epitomized the new absolute frame of reference—with clear before and after attributes and a continuous linear flow of motion. The religion of this period also complemented this mechanistic analogy. God was the master mechanic or clock-maker, and the universe was essentially wound up and allowed to tick on its merry way. Industrialization of the modern world also involved machines. The unchanging cycle of an agricultural society was replaced with the linear progression of assembly lines. Newton enthusiastically endorsed the machine metaphor and made it the root metaphor of his scientific method and view of the world.

Although the dominance of this method in classical physics is indisputable, this view of the world is not a timeless truth. It is merely one approach in a long history of method that includes several approaches which do not depend upon linear time at all. Indeed, Newtonian mechanics was eventually itself rejected by many scientists (see Chapters 1 and 10). Developments in physics—including relativity and quantum theories—have challenged Newton's conception of time and with it the five characteristics of explanation: objectivity, continuity, linearity, reductionism, and universality. When the absolute character of time was questioned, absolute motion and the entire Cartesian approach to method was also challenged.

The Newtonian Framework of Method

On the whole, however, this modification of method in physics has not translated into a modification of method in psychology. Although psychology has been liberalized in some ways with the advent of cognitive approaches (see Chapter 5), psychology's dependence upon a Newtonian framework for science has not significantly diminished. Psychology has become more permissive than Newton on matters of theory,[15] but all five characteristics of his framework are clearly evidenced in mainstream method.

Objectivity

To begin with, the temporal aspects of method are presumed to be objective. Time is an independent medium in which methodological events must, of necessity, take place.[16] It is thus the purpose of any methodology to track and measure these events as they occur across time. Experimental variables are thought to have an objective duration (which is measured by timepieces) as well as an objective sequence. Particularly important here, of course, is causal sequence (or the first in any "chain" of events). Causal processes are considered to take place in a particular temporal order, and "fill up" the time interval between the onset of cause and the onset of effect. Psychological method is arranged to take into account these temporal properties of causal processes. For example, it is widely believed that independent variables must precede dependent variables in experimental design.

Another important sign of time's objectivity is its anonymity. That is, the notion that linear time is "real" leads to its obscurity in method. Despite the obvious pervasiveness of temporal factors at all levels of method, few of these factors have received systematic attention. As McGrath and Kelly note,

> We often use delays, latencies, and rate and duration measures; but these are almost always considered as ways to measure or control other variables, not as parameters interesting in their own right. Virtually every experiment is temporally bounded. . . . Yet the assumptions and limitations involved in this time boundedness are seldom noted.[17]

Because linear time (like space) is an objective medium through which psychological events must pass, it is not thought to affect those events except in predictable and knowable ways. Hence, time's familiar existence "in reality" makes it uninteresting and unworthy of study.

Continuity

Other aspects of Newton's absolute time contribute to its lack of recognition in method. The fact that time is viewed as continuous is also contributive. Continuity in this sense implies two somewhat overlapping properties of time: consistency and uniformity. A consistent medium provides events (in time) with predictability, whereas a uniform medium allows temporal processes to be invariant (from one period of time to the next). Given time's continuity in these two senses, it is no wonder that time has never captured the interest of methodologists. *Linear time is a "variable" without any variance. Quantity* of time can be varied, but *quality* of time is presumed to be constant in every way.

The word "presumed" is important here because psychological researchers have rarely *investigated* time's quality in relation to a subject of interest. Indeed, the possibility of variance in the quality of time has rarely been considered. Objective time is assumed to be unaffected by the events that supposedly "fill" it, and researchers routinely assume complete equality between two separate, but equal, quantities of time. Current conceptions of replication, cross-validation, and test reliability depend upon this assumption.[18] If an experiment is replicated, for instance, the measurement or control of time's quality is never considered. It is assumed that linear time needs control only in the quantitative sense. Time cannot accelerate or slow down; time cannot vary with the processes under study; and time certainly cannot jump unpredictably from one dimension or scale into another. If time did alter its quality in any of these ways, no two experiments or tests could be meaningfully compared, at least as conventionally considered.

Interestingly, though, some psychological researchers report the existence of discontinuous phenomena. As described in Chapter 2, for example, many developmental psychologists view stages of development in a discontinuous and sometimes unpredictable manner. Transition between stages may occur abruptly, without the graduated steps between stages that are expected of a continuous framework. Life-span psychologists also contend that some early forms of development are inconsistent with later forms.[19] Many later patterns of behavior appear to emerge without any continuity to prior development.[20] The methodological problem with such phenomena is that they are not uniform across time. They are, in a real sense, unpredictable, and thus difficult to investigate with methods that assume continuity.[21]

Linearity

This difficulty can be further elaborated through another aspect of Newton's temporal framework—its linearity. Because events are viewed as happening along a "line of time," the metaphorical properties of a line are incorporated into our research designs. For instance, conventional experimental design calls for only two observations—pre and post. Although multiple groups of experimental and control subjects may be used, only two observations—across some specified "length" of time—are considered necessary for inferences about the process under study. The reason is that the processes are themselves thought to be linear. They are viewed as being "pushed" by the earlier events and remaining smooth (and continuous) like a line. Therefore, a comparison between two measurements on the line of events is all that is necessary to indicate the pattern of a linear "flow," and thus the nature of any future events in this flow.

As a contrast, consider the measurement of so-called "nonlinear" phenomena. Some scientists hold, for instance, that human life is filled with cyclical phenomena—from diurnal rhythms to the oscillations of seasons.[22] Pre- and postmeasure methodologies are not only insufficient for such phenomena, they are misleading. Pre- and postmeasures could be taken at any point in the ebbing and peaking of the phenomenon, making data interpretation virtually impossible. Indeed, if important nonlinear processes do exist in psychology, *none* of the so-called "true experimental" designs would be adequate to assess them.[23] All are based on some variant of pre/post linearity.[24] Cyclical processes would require a series of observations distributed systematically across a relatively long (and theoretically specified) "span" of time. Interestingly, this type of design is lower in the methodological pecking order: "quasi-experimental" rather than "experimental" in status.[25] This means, of course, that researchers are relegated to a less controlled and lower status method for assessing nonlinear phenomena.

Methodologists Campbell and Stanley appear to sympathize with this. After listing all the problems with quasi-experimental designs, they offer these words of solace:

> This formidable list of sources of invalidity might, with even more likelihood, reduce willingness to undertake quasi-experimental designs, designs in which from the very outset it can be seen that full experimental control is lacking. Such an effect would be the opposite of what is intended. From the standpoint of the final interpretation of an experiment and the attempt to fit it into the developing science, every experiment is imperfect.[26]

Statements like these may seem reassuring, but moral support is rarely helpful to the researcher who is attempting to publish experimentation in which there are inherent problems with controls. The usual tack is to work primarily (if not exclusively) with linear processes because they submit themselves to "tighter" designs (and a better publication record). The net effect, of course, is that linear phenomena are given the most visibility and credence. The problem is that this visibility and credence is an artifact of a pecking order which is itself based upon linear assumptions.

Linear assumptions are also implicated in the common methodological expectation that change is linear. That is, patterns of change are thought to unfold in the same linear fashion as time itself—"longer" is assumed to be proportionately "more." Indeed, time is routinely viewed as having ratio (or interval) properties of measurement, though this is seldom explicitly stated.[27] If fifty units of change have occurred in ten minutes of "objective" time, we are apt to expect twice that amount of change in twice that amount of time (everything else being equal). It is as if we expect linear time to map isomorphically to all other scales of measurement. The ratio properties of timepieces are presumed to endow other measures with similar properties, yet such presumptions have never been tested scientifically.

Universality

The uniformity of this temporal "scale of measurement" is the basis for another Newtonian characteristic—universality. This may well be the most criticized of all aspects of Newton's legacy for psychological method.[28] Universality is the notion that scientific method is intended to discover laws that are timeless in the sense of unchangeable and eternal. Although each part of the lawful process must take place across time in the objectively continuous and linear manner described, the process-as-a-whole is independent of the particular temporal context in which it is detected. "Principles of conditioning" and variants such as Skinner's "schedules of reinforcement" supposedly occur without regard to the historical or cultural context in which they take place.[29] If anything, laws like these are thought to *produce* historical and cultural contexts rather than result from them. Laws of reinforcement, for instance, cause societal movements rather than become an effect of them.

As described in the history lesson above, laws are thought to lie "behind" the temporality and change of the world as perceived and are the ultimate scientific truths. Method is the means by which the scientist gets beyond superficialities, such as cultural context. In fact, method considers such superficialities "extraneous variables" that require control rather than examination. Because universal laws stem from relationships that transcend the present, factors related to the present context of the data are extraneous.

This provides a vital part of the justification for laboratory experimentation and computer simulation.[30] These methods deliberately filter out natural and contextual factors because these presumably hinder study of the more basic processes. Temporal context can be "added in" to these methods, but what is included is ultimately a model that does not occur in the "real time" of the process under study—i.e., the natural sequence and directionality of events.

A potential problem is that some psychological processes may be understood only in their natural temporal contexts. In cognitive psychology, for instance, Neisser discusses the phenomenon of "everyday memory" in this manner.[31] He wonders why cognitive researchers have so rarely investigated this and doubts whether the usual laboratory approaches can examine it properly.[32] Part of his concern seems to be the investigation of context-bound processes by contextless methods.[33] Dreyfus appears to have analogous concerns regarding artificial intelligence.[34] He insists that computer simulations of our intelligence cannot capture the contextuality and situatedness of the human mind. Computer simulations are universal in their logic, and thus must have context "modeled into" them. Dreyfus attempts to demonstrate that rule-governed machines, such as computers, cannot incorporate (or represent) the intricate and changing context of the human mind (see Chapter 5).

This is tantamount to saying that some psychological processes are *not lawful* in the Newtonian sense. If rules and laws cannot be divorced from their temporal contexts, then these so-called laws do not apply to the same processes in different times and spaces. Gergen attempts to make this case in social psychology. He contends that social psychology has failed to establish lawful relations because such lawful relations do not exist. The subject matter of social psychology is contextual and nonuniversal by its very nature:

> The terms in which the world is understood are social artifacts, products of historically situated interchanges among people. From the constructionist position the process of understanding is not automatically driven by the forces of nature, but is the result of an active, cooperative enterprise of persons in relationships.[35]

Cultural and social influences are so strong that *they* govern the very fundament of social relations. Social psychology is thus inherently situated in a historical and cultural context rather than a by-product of lawful principles that are independent of such influences.

Reductivity

Newton's final characteristic of explanation—reductivity—has long been considered an aspect of good methodology. A scientist must reduce complex problems to their constituent parts and study those parts before they are reassembled. Still, the prominent role of linear time in this procedure has not been explicit. Actually, absolute time *requires* a reduction of processes because the process-as-a-whole never exists at any one point in time (see Chapter 1). If processes occur along the line of time, only a piece of the process is available for study at any point in time. This implies that the interactions between components of the process (two or more points in time) can only be studied through recording devices (or the memory of the scientist) because any preceding component is no longer in view.

This distinction may seem trivial at first, but it has momentous implications for method. The most important is that *whole processes are never directly observable* from a Newtonian perspective. Metaphoric "photographs" of each moment of the process can be "taken," but then only the photographs can be examined. Scientists have no direct access to the *interactions between component parts*. This is no small implication to positivistic scientists who pride themselves in studying "observables" only. Such a restriction forces them into a reductionism, either to parts of the process—without direct access to their relationships across time—or to the possible distortions of such relationships through recording devices.

The emphasis on across-time processes also leads to the neglect of simultaneous processes. When relations across time are the focus, relations "within" time are given short shrift. As noted in the previous chapter, several personality theorists regard contemporaneous processes as crucial. Moreover, many systems theorists and family therapists argue for the existence of holistic processes which require synchronous events (see Chapter 8). They point to systemic forces that seem to operate more as a gestalt of concurrent events rather than a linear cause from the distant or immediate past.[36] This is not to say that causal analyses are eliminated. A nonsequential form of causation can be substituted, referred to variously as "ahistoric,"[37] "formal,"[38] or "synchronous"[39] causality.

Alternate approaches to causation begin to hint at alternate approaches to method. As dominant as the Newtonian approach has been in psychology, its inability to handle nonlinear phenomena has spawned several "anomalies." Whole methodologies have been formulated which serve as radical alternatives to mainstream methods in psychology. Space constraints prohibit a thorough review of them here. However, two of the more important and exemplary regarding time are outlined: one originally from the natural sciences and one originally from the humanities. Although both endorse some form of time in their assumptions, each has its own way of rejecting Newton's particular temporal framework for explanation.

Anomalous Methods

Systemic Method

The first method is sometimes referred to as "systemic" or "structural" in its procedures.[40] Its original impetus was the conviction that complex objects or systems should be dealt with as wholes. Although this notion has been around for centuries, its formal proposal as a method occurred in biology through the study of organic systems. The biologist, Ludwig von Bertalanffy, was the early architect of this "organismic" viewpoint.[41] However, his conceptions were clearly underlaid by those of the philosopher, G. W. F. Hegel.[42] Hegel held that the identity of any particular thing—what the thing *is*—stems from its relation to the whole. That is, many of the qualities of things are not *in* the things themselves; they are qualities stemming from their *relations* among other things.

The notion of *emergence* is critical in this regard. As described in Chapter 1, a molecule of water is an emergence of two atoms of hydrogen and one atom of oxygen. Studying any of the atoms in isolation of the water molecule does not permit us to predict the qualities of water when the atoms are combined. The qualities of this compound are only discernible once the whole of the water molecule has "emerged." This means that the process of emergence is a *discontinuous* one, by definition. At one point in time the entity is a set of individual hydrogen and oxygen atoms with their respective physical properties, and at the next point in time the entity is a molecule of water with its qualitatively different properties. A holistic and qualitative change has occurred—from one whole to another—that cannot be explained (or predicted) through piece-by-piece gradual change. This qualitative change is predictable only because we have observed this emergence happening under certain conditions, not because previous states predict future states.

The phenomenon of emergence also raises questions about conventional *linear* conceptions of causality. The Newtonian notion that a prior event determines a later event does not explain emergent phenomena. As James and Robert Haldane wrote in 1883:

> It would thus appear that the parts of an organism cannot be considered simply as so many independent units, which happen to be aggregated in a system in which each *determines* the other. It is, on the contrary, the essential feature of each part that it is a member of an ideal whole, which can only be defined by saying that it realized itself in its parts, and that the parts are only what they are insofar as they realize it.[43]

In this sense, each part of an organic system is "caused" by its *present* relation to the other parts, and not its *past*. The qualities of the water molecule do not stem from relations *across* time; they stem from relations *within* time (at the same time), as parts interact in the present and every present moment thereafter. Change, then, is from one discrete whole to the next, without one being causally chained or "determined" in the classical sense.

Bertalanffy found Newtonian methods wanting for other reasons as well. First, mechanistic methods are insufficient to investigate open systems. Closed systems are isolated from their environments and thus predictable from their initial conditions. Living systems, on the other hand, are open systems. Open systems exchange information or substances with their environments and are not, therefore, predictable from their initial conditions. Although closed systems inevitably move to a disordered state (entropy), he felt that open systems could *increase* their order over time. This means that the final condition of an open system cannot be predicted from its initial condition, and so simple linear causality does not suffice.

Bertalanffy also felt that open systems evidenced another property foreign to linear accounts—*equifinality*. Equifinality postulates that some systems have goals of their own, a final objective toward which the system is striving. Systems with this quasi-purposive quality may reach the same state from qualitatively different initial conditions. It is almost as if the *future* were as influential as the present or past. The linear explanations of Newton clearly do not take this natural teleology into account, nor do Newtonian explanations seem to handle the holistic properties of open systems. As Bertalanffy stated it:

> Mechanism . . . provides us with no grasp of the specific characteristics of organisms, of the organization of organic processes among one another, of organic "wholeness," [or] of the

problem of the origin of organic "teleology." . . . We must therefore try to establish a new standpoint which—as opposed to
mechanism—takes account of organic wholeness, but . . . treats
it in a manner which admits of scientific investigation.[44]

So far, then, systems methodologists have been shown to violate three of the five characteristics of a Newtonian framework for
method: linearity, continuity, and reductionism. However, their movement away from the other two characteristics—objectivity and universality—has been less consistent. All systems methodologists give
these two characteristics less weight than Newton (and mainstream
psychological researchers). Few, for example, do laboratory or simulation studies because the natural context of the system (itself part
of the whole) is lost in these methods. Bertalanffy, though, felt that
it is vital to study the *general* properties of all systems—whether
they be physical or social in nature. He contended that there is a
general and ideal form for all such systems. The moment by moment conditions of an individual system are important, but there is
also a timeless form worthy of study. In this sense, Bertalanffy's
proposed method does not completely avoid the old Cartesian (and
Newtonian) goal of universal and objective knowledge. Some truths
are still viewed as systemic "laws" that hold for any system, regardless of its situational context.

Other systemic methodologists have thoroughly repudiated the
universality and objectivity of a Newtonian framework. The work
of linguist, Ferdinand de Saussure, is perhaps the best example of
this. Saussure abandoned the atomism of classical grammar, stressing instead the underlying system of relations within language for
meaning. He felt that words cannot mean in isolation. Words re-
ceive their meanings from their relations to other words in the
same system. Similar to Bertalanffy and Hegel, Saussure saw systems of meaning as discontinuous, nonlinear, and holistic in character. Meanings are derived from their synchronous relations to
one another and thus change discontinuously and noncausally when
viewed across time. As Saussure himself put it, "language is a
system of interdependent terms in which the value of each term
results solely from the *simultaneous* presence of the others."[45]

Unlike Bertalanffy, however, Saussure went on to assert that
these properties affect not only superficial structures but also the
supposedly deeper and more general structures. Before Saussure,
linguistic meanings were studied as evolutionary (and linear) phenomena—the meanings of words were considered to evolve from
earlier meanings and language forms. Saussure called such meth-

ods diachronic methods, because they assume that across-time relations are the most crucial to meaning. Saussure, however, advocated synchronic methods that focus upon language as it stands in the present. These methods take a cross-sectional view, examining relations among the various concurrent parts of language. In this sense, underlying linguistic relations do not necessarily have a general systemic character that transcends the synchronic present. There is no fixed set of systemic relations to which languages must conform (as Bertalanffy advocated about general systems). Linguistic concepts have an important arbitrariness to them, deriving their meanings through convention, and not through independent and universal structure.

In conclusion, the approaches of Bertalanffy and Saussure serve as illustrations of the assumptions and uses of systems methodology. They differ greatly from conventional Newtonian methods in psychology. They do not reduce the phenomena under study—at least not until they have a sense of the qualities derived from their relationship to the whole. They look less to the past for the cause or explanation. They are more likely to look toward the synchronic present and attempt to answer "why" questions by locating the event within the larger system rather than the singular past. They do not necessarily expect change to be continuous or proportionate to the quantity of objective time. Therefore, conventional pre/post designs are not as useful and are rarely employed.

Hypothesis testing and rigorous control are also less pertinent, because the context of the subject matter is more heavily emphasized. Though systemic methodologists differ to some extent in how much context is stressed, all value the natural descriptive techniques of field studies over laboratory observation. The structuralist Piaget is an excellent example of this.[46] Most of his data were developed from his own natural and informal observation of children. Unlike Newtonian universal methods, Piaget sometimes followed his own intuition when he was interviewing children. He did not feel constrained to standardized formats, or to formally logical or "mathematical" theorizing. He also did not restrict his interpretations of change to the continuous variety, freely observing discontinuities in children as they developed.[47]

Hermeneutic Method

Unlike systemic methods, hermeneutic methods were originally derived from the humanities rather than the natural sciences. The term "hermeneutic" comes from the Greek word *hermēneuein* ("to interpret"). Some writers have held that the word is related to Hermes, the messenger of the gods in Greek mythology. Just as Hermes supposedly helped humans understand the meanings of godly messages, hermeneutical method attempts to help the researcher understand the meaning of any text or behavior. The first formal use of this method seems to have been in religion. Ancient biblical scholars were vitally interested in the original meaning and intent of the various authors of the Bible. Consequently, they developed principles of interpretation to facilitate this endeavor. These principles were later extended to other texts of classical antiquity as well as legal cases during the Renaissance period, giving impetus to the expansion of hermeneutic procedures into a general interpretive methodology.

Wilhelm Dilthey is the person normally credited with moving hermeneutics to the data of the human sciences.[48] Because the meaning of human actions is not always apparent, he felt interpretive techniques are necessary to provide them with intelligibility. Dilthey took the hermeneutic procedures that successfully produced an understanding of ancient texts and legal cases, and applied them to the human realm of both verbal and nonverbal relations. His method was not unlike that of Saussure. He was interested in the "structural coherence" of social organizations and particularly attended to "expressions of life," such as art and literature. An understanding of human experience was possible "because between people and a state, between believers and a church, between scientific life and the university there stands a relation in which a general outlook and unitary form of life find a structural coherence in which they express themselves."[49]

Dilthey investigated this structural coherence through the whole-part relations of the phenomenon being interpreted. Part of this phenomenal whole is the interpreter because the interpreter has expectations and cognitive biases that affect interpretation of the phenomenon, just as the phenomenon itself affects the interpretation. In this sense, the entire method has a circular character. It "begins" with a preliminary perception of the whole.[50] This is analogous to the vague apprehension one must have to ask a question in the first place. The process then moves to the meaning of the whole, and to a change in the sense of this meaning as one's awareness of the phenomenal details grows. During this back-and-forth movement between interpret*er* and interpret*ed*, understanding of the phenomenon is deepened. Dilthey called this the "hermeneutic circle," and it remains the heart of hermeneutic methodology.[51]

Dilthey felt, however, that the human realm can never be fully known, at least not in the universal sense. Its whole-part relations are perpetually in process, with the tools of knowledge themselves in process. Unlike the universality of Newton, Dilthey held that there is no final perspective outside the subject matter from which to view the totality of life. Such things as social life, poetry, and art are vital to a person's life, and there is no once-and-for-all understanding of their lived meanings. There is no such thing as *the* interpretation. Interpretation is always interactive— between the interpreter and the text—and ever-changing within time. Thus, Dilthey's contentions are the epitome of a contextual perspective.

This perspective on method is typically quite provocative to psychologists.[52] Of course, it challenges all the Newtonian characteristics of method, but its explicit attack upon universality may be the most difficult for mainstream psychologists to accept. The reason is that universality goes right to the heart of what method is supposed to accomplish—advancement of knowledge. Knowledge is viewed as being universal, by definition. Knowledgeable explanations cannot change from one moment to the next; they must cut across the conditions of particular moments and focus upon what is universal in them. Otherwise, they are just temporary "understandings," and only benefit those in the particular moment in which the understanding is gained. Real knowledge, on the other hand, is thought to transcend the particulars of the moment, so that it can transfer to other temporal contexts.

What is often not acknowledged in this familiar approach to knowledge is the need for absolute time. Without the consistency and uniformity of absolute time, this approach to knowledge would not be possible. No piece of knowledge could ever be counted on to transfer from one moment to the next. Processes would change as the quality or scale of time changed. Just as a melody would be different when played to a different rhythm, so knowledge would seem to lack unity when occurring in different time frames. The otherwise steady beat of a healthy heart could appear erratic or stopped completely with a change in the quality of time. Although no hermeneuticists are contending this, it illustrates how temporal assumptions permeate our most basic assumptions of knowledge and method.

The philosophy of Martin Heidegger is particularly pertinent on this point and basically underlies the hermeneutic method. Heidegger completely rejected the Newtonian notion of absolute time, contending that all knowledge is *contextual*.[53] The term "contextual" is meant to connote both temporal and spatial "conditions." From a Newtonian perspective, these conditions of context include three dimensions of space and one dimension of time. From a Heideggerian perspective, however, lived time and space *both* include three dimensions, even when speaking only of the "now." That is, all three dimensions of time—the past, present, and future—*co-occur*, just as all three dimensions of any spatial situation co-occur. The past as memory and givenness, and the future as anticipation and possibility, exist in and provide a vital simultaneous context for the present.

Newton, of course, postulated a linear sequence of temporal dimensions, with the past, present, and future forever separated in time.[54] Each dimension is encountered in its turn, but no dimension can be the simultaneous context for any other dimension. Universal laws are thus needed to cut across these separate dimensions and provide knowledge that is applicable to more than one point in time. However, from Heidegger's perspective, universals accomplish this at the expense of the contextual present.[55] Instead of valuing the unique and context-filled present, universals strip moments of their individual uniquenesses and relate together only their similarities. What is left is an impoverished and inapplicable aggregate of moments. The aggregate cannot be applied because it has nothing to do with the unique context of any particular moment, only the contextless aggregate of moments.

All is not lost, however. According to Heidegger, the recognition that time's three dimensions are simultaneous, rather than successive, permits any particular moment to be itself a temporal whole. Knowledge of the present, in this sense, is not just knowledge of one point on a one-dimensional line. It is an integrated knowledge of all three temporal dimensions. A nonlinear type of universality is preserved because a *full* understanding of the now implies some understanding of the past (memories) and future (possibilities) as well. The question is, how does one attain this full hermeneutical understanding? How does one construct a method on such radical notions? Heidegger's contextual assumptions challenge the whole methodological enterprise of seeking universal laws through contextless methods.

One of Heidegger's students, Hans-Georg Gadamer, has probably been the most responsible for providing the methodological implications of Heidegger's philosophy.[56] Gadamer notes that one of the primary functions of method is finding the truth about the world. Consequently, most methods operate on the principle of correspondence: Does the concept under consideration *correspond* to the truth of reality? This is the logic of mainstream psychological method. Researchers test their hypotheses or predictions to see if they correspond with their data, and statistics are essentially intended as a means of gauging this correspondence. An objective method is thought to be one that results in the closest correspondence with universal truths or empirical laws. The only problem, given Heidegger's assumptions, is that there is no universal truth to which concepts can correspond.

This might seem to leave us without a method, and it does in the conventional sense. Gadamer, however, asks whether a *contextual* method is possible. Certainly, this type of method would require another type of truth, a contextual truth that does not exist "outside" the events to which it pertains, but a truth that "coheres" in the events themselves.[57] His affirmative answer to this question is initially difficult to grasp because truth (like knowledge) has become so synonymous with universality (and thus absolute time). Truth is thought to be "independent of consciousness" (objective) and "for all time" (universal).

Contextual truth clearly cannot be objective and universal in this way. Nevertheless, it can have its own "objectivity" and "universality" in a contextual sense. Contextual truth can be objective in the sense of affirmed and shared by others. Although it does not exist in a reality apart from consciousness, it nonetheless exists in the experiences of humans as they articulate it. For example, facts in this sense are the shared meanings of people involved in the contexts, not absolute data points in reality. Contextual truth can also be universal, after a fashion. As noted above, the integrated whole of time permits truth to have a nonlinear type of relevance for all of "time." It is also important to realize that this type of truth is manifested in *how* things are rather than *what* they are. The how is located in the world as an ongoing process, not in the world as "is." In this manner, one can legitimately ask questions in one of these contexts and through a method obtain answers that can be right or wrong.

Faulconer and Williams illustrate this with a chess game.[58] When a chess master is asked for the appropriate move, the correct answer to this question is not one that corresponds to a universal (or ideal) chess game. The correct answer is one that considers the uniqueness of the game as a whole, as represented by the particular configuration of the pieces at the moment. This present configuration includes both the prior movements of the pieces (the givenness of the past) and the possibility of piece movements (the prospect of the future). What the question of "appropriate move" seeks is an *understanding* of the present situation. That is, given what is going on—the context of the game—certain questions arise and only certain answers can address the temporal context "correctly." There are clearly right and wrong answers to the question, and yet there is no absolute ideal, no universal truth against which to compare the answer. Its correctness is governed by the context of the game itself.[59]

According to hermeneuticists, then, psychologists should seek situated understandings and not universal laws. Psychologists cannot avoid being context-bound either in their methods or in their objects of investigation. Analogous to the chess game, psychologists are player-participants in the midst of a temporal game in their research. The game is an ongoing process from which the researcher cannot be removed without dramatically altering (Heidegger would say destroying) the subject under investigation. This does not mean that researchers are resigned to subjective relativism. Intersubjectively agreed upon meanings have always been the hallmark of science,[60] and right and wrong interpretations can be discerned. Nonetheless, the criteria for their rightness are not outside the context of the interpretive situation itself.

Conclusion

The fact that there are alternative methods helps us to see that traditional methods are not without metaphysical assumptions. Traditional methods are not neutral tools of inquiry, discovering objective laws of reality. They have, like all methods, very definite assumptions about their subject matter and the nature of inquiry. The supposed objectivity of the Newtonian method, however, has obscured the assumptions which undergirded it. As the philosopher Edwin Burtt put it, Newton "made a metaphysics of his method."[61] Newton imbued his method with a metaphysic that prevented him from seeing anything but his metaphysic in his empirical findings. He saw absolute temporal relationships between the variables under study because his method would not permit him to do otherwise.

Mainstream psychological research appears to be in a similar state. Like Newton, many psychologists seem blind to the metaphysics inherent in their method. Implicit are all five of the temporal characteristics of the Newtonian method. Measures of time and chronological order are examples of the many *objective* factors presumed to be involved in empirical findings and research design. Research practices based on replicability and predictability assume that time is like space—*continuous* and variant only in quantity and not quality. *Linear* assumptions permeate psychology's best research strategies, from pre/post designs to the pervasive expectation of linear change. The purpose of these designs is to *reduce* complex processes to their components—as separated in linear time—so that *universal* and context-free laws can be discerned.

Many critics of the Newtonian tradition focus upon phenomena that are less amenable to this method. Some claim that there are psychological processes which are subjective, discontinuous, nonlinear, holistic, and/or context-bound. As a result, whole methods have been developed to understand these processes better. These anomalous methods make dramatically different assumptions—for the most part abandoning the Newtonian metaphysic of time. Change and even direction of change are still endorsed, but Newton's assumptions about time's absolutivity and independence from context and perceiver are generally avoided (see Chapter 10).

Two alternative methods are specifically reviewed, one originally adapted from the natural sciences and one originally adapted from the humanities. Systems methods are formulated specifically to examine holistic and discontinuous processes. Rather than the usual cause and effect sequences, systemic methods rely upon synchronous causality or relations that are within instead of across time. Hermeneutic methods, on the other hand, are devised to tackle the interpretive and contextual elements of scientific investigation. One of the cornerstones of this approach is the notion that all interpretation is a circle, including the observer as well as the observed. Knowledge in this sense is not absolute and unchanging but inextricably tied to the meaningful moment in which it is needed.

Chapter 5

COGNITIVE PSYCHOLOGY

Cognitive psychology seems to be the centerpiece of modern psychology. Cognitive explanations have come to dominate not only the basic aspects of the discipline, such as learning and language, but also the more applied aspects of the discipline, such as educational psychology and psychotherapy. Before the cognitive movement, psychologists restricted themselves to behavioral explanations. The early Newtonian focus upon observables meant that observable behaviors (e.g., responses) had to be explained in terms of observable environments (e.g., stimuli). The importance of mental processes was either denied or downplayed. The rise of cognitive psychology, however, meant that explanations could include the mind—first as a passive filter of environmental stimuli and later as an active mediator and processor of information.

Many observers have claimed that such cognitive explanations are nothing short of revolutionary.[1] The question is: has this so-called revolution affected psychology's Newtonian paradigm? Clearly, Newton's positivism and explanatory style are largely responsible for psychology's *pre*cognitive method and theory. Not only were observables emphasized, but nearly all explanations presumed absolute temporal characteristics. The relation between environment and behavior was almost exclusively portrayed in a linear sequence. Stimulus and response, for example, occurred in sequences across time, with stimuli and conditioning histories given primacy in explaining present responses.

Have these linear aspects of psychology's metaphysic been revolutionized with the appearance of cognitive explanations? This chapter reveals that they have not. Although mainstream cognitive psychologists have been willing to theorize about nonobservables, they have not only maintained their Newtonian heritage, their models have been instrumental in *preserving* the Newtonian paradigm for modern use. One of the main reasons for this is the preservation of Newton's mechanism. Newton had brilliantly coalesced the five characteristics of his temporal framework for explanation as a machine metaphor (see Chapter 1). The universe supposedly functioned as a great clock, flowing continuously and objectively along the line of time.

As this review shows, mainstream cognitivists have not fundamentally altered this explanatory metaphor. In fact, they have updated it to the modern digital *computer*. Although this updated metaphor has placed more emphasis upon the unobserved "software" than in behavioral accounts, it has not changed the linear characteristics of the machinery. Indeed, the computer has provided a modern means by which the past is preserved for later influence in the present. Computers are viewed as excellent *stores* of information. The human mind is often considered to be analogous to this information-processing machine through its memory capacities. Even when the computer analogy is not obvious, mainstream models have presumed some sort of linearity of the environment, either through the distant past (memory) or the immediate past (input). The past still retains its primacy.

These mechanistic metaphors have not gone unchallenged. Outside mainstream cognitive psychology, the analogy with a computer's sequence and storage functions has not only been questioned, but the five Newtonian characteristics of absolute time have also been disputed. This chapter examines each of these challenges, particularly as they relate to time and explanation. First, however, it is important to understand the history of cognitive studies in psychology. How did this field come to rely upon absolute time, and what are the historical antecedents of the anomalies that challenge this reliance?

Time and Associationism—A Brief History

Most cognitive psychology texts trace the ancestry of current ideas to Aristotle.[2] Although many disciplines can probably be traced to Aristotle, it is clear that his philosophy makes unique contributions to the metaphysics and epistemology of modern cognitive psychology. Regarding metaphysics, Aristotle's view of time is in many ways similar to Newton's.[3] Aristotle recognized the importance of cyclical time but, like Newton, believed that time exists independently of events.[4] Just as absolute space is independent of consciousness and uniform, so time is viewed as objective, continuous, and universal (see Chapter 1). Space gives an objective structure to the world of things, and time gives an objective structure to the world of events.

Aristotle's brilliance and philosophical consistency led him to incorporate these aspects of his metaphysics into his epistemology. Because events of the world supposedly occur *in* time, any knowledge of these events must take this into account. This implied for Aristotle three fundamental epistemological principles that continue to play important roles in contemporary cognitive psychology: empiricism, contiguity, and repetition. *Empiricism* in this context is the notion that knowledge is derived—at least in large measure—from our experience of world events as they are organized separately from us. The structure of these events in time and space exists apart from us, and must be correctly perceived to be correctly understood.

The manner in which events are temporally structured concerns the other two epistemological principles—contiguity and repetition. *Contiguity* was considered one of several "laws" of thinking by Aristotle.[5] This particular law postulates that events vary in their contiguity to one another in time. Those events that are more contiguous or near to one another in time (and space) are more likely to be associated. Events that are farther apart in time (and space) are less likely to be associated. *Repetition* was not a formal "law" of thinking but a vital part of Aristotle's epistemology.[6] Repetition is the notion that events which occur frequently across time are those likely to be learned. Events that occur only one time may be learned, but these are not learned nearly as well (and may not be as important) as events that repeatedly present themselves.

The significance of Aristotle's metaphysics of time for these three epistemological principles is often missed. Without an objective and spatialized view of time, each principle is undermined. Empiricism, for example, requires that time be a property of reality so that it can structure our experiences. Otherwise, time would be in our minds molding our experiences, rather than in our reality molding our minds (see Chapter 3). Contiguity also requires an objective framework for time. The chronological "nearness" of events assumes that some "spatial" frame of reference is available for comparing temporal "distances." The same is true for repetition. The frequency of an event's occurrence connotes some sequencing across time as well as intervals of measurable time, independent of the events themselves.

The combination of these principles has become known as *associationism*. Working in the tradition established by John Locke (see Chapters 2 and 3), Thomas Hobbes, George Berkeley, and David Hume articulated the associationist outlook in its most extreme form. They contended that human knowledge consists of associations of simple ideas. Simple ideas, such as redness, roundness, and sweetness, are sensed through experience, and then combined through their contiguous and repetitious relations into more complex ideas, such as that of an apple. This meant, of course, that complex ideas could be broken into their constituent parts for analysis. Although some in this tradition specifically fought this implication (e.g., John Stuart Mill), the majority of associationists held to some variation of this reductionistic theme. As we shall see, some of these variations have continued in cognitive psychology to this day.

Certainly, associationism is the foundation of traditional conditioning theory. Whether it be classical or operant conditioning, the events to be conditioned must be experienced directly (empiricism), presented together in time (contiguity), and occur over several trials (repetition). In classical conditioning, the unconditioned and conditioned stimuli must be repeatedly presented together in time, and in operant conditioning, the reward and targeted behavior must be similarly contiguous and repetitive. A Newtonian framework for time is implicit in these procedures. For instance, the events of each conditioning operation are viewed as objectively and reliably contiguous in the environment. If time were subjective, then the contiguity and repetitiveness of the events would be within the perceptual control of the subjects, rather than the environment.

The German scholar, Hermann Ebbinghaus, is perhaps most responsible for applying associationist assumptions to cognitive research. His famous work with nonsense syllables (e.g., zok and luj) was predicated on the associationist notion that people would already have associations with real words through their prior experiences. He reasoned that one had to have materials which had no past associations to understand how associations were *first* formed. Ebbinghaus was also one of the first cognitive researchers to introduce time factors into his methodology. He carefully controlled quantities of time in his experiments and was vitally interested in the rates at which learning occurred. He agreed that contiguity and repetition were the fundamental determinants of memory.[7] The more often he rehearsed (or repeated) contiguous lists of nonsense syllables, the greater was his memory for these associations (the first learning curve[8]).

Early cognitive researchers in the United States extended Ebbinghaus's pioneering efforts. Although topics and methods changed, these researchers retained his essential associationist theory of memory. For instance, the role of *meaningfulness* in memory was a central concern in the 1950s and 1960s. True to associationism, the most meaningful items were considered to be those items that the person had: (1) the most contiguous associations—e.g., the greatest quantity of associated words,[9] or (2) the most repeated exposures—e.g., the most familiarity.[10] Interest in contiguity was so high that methods were developed to manipulate contiguity. Subjects learned words in various pairings to discern how such associations affected one another, eventually resulting in several important principles of forgetting. In fact, this type of research was so successful that many cognitivists of the late 1950s and early 1960s believed that all the fundamental laws of thinking could be established through associationist analyses.[11]

Still, dissenting voices were heard at this time. Their primary concern was the mind's passive nature when considered from this associationist perspective. Because time existed outside the mind— as a primary determinant of associations and meaningfulness— environmental events were viewed as the active organizer of cognition. Communications engineering and information theory, however, led to research showing that humans *coded* the associations (or information) they received. Although the associations were themselves preserved in the code, the mind actively converted the information from one form into another. Research became centered less on the content of the associations and more on the cognitive processes which acted upon the associations.

How were such processes thought to operate? Like all researchers, cognitivists needed tentative models to generate hypotheses for experimental tests. Since the time of Newton and before, mechanistic models have guided hypothesis generation. For instance, some ancient peoples analogized the mind to a machine of their day—the catapult[12]—as Freud analogized the mind to a machine of his day— the steam engine.[13] Latter-day cognitivists take a similar tack in this instance. They analogize the processes that act upon the associations (now called "information") to the digital computer. The computer is a system for processing symbolic information. The computer still depends on the environment structuring information when it is encoded (e.g., through contiguity). Nonetheless, the computer is also guided by previously encoded associations, called a program, that have been stored and preserved in its memory. Stored information can be retrieved through feedback operations and act upon new information.

Perhaps the pinnacle of the computer metaphor is the multistore model of memory. This model dominated cognitive psychology during the 1970s and early 1980s and still figures prominently in many recent cognitive texts.[14] In this model, external stimuli (associations) are processed in several stages across time, each stage constituting a different type of memory storage. Note again the reliance upon time here. Each stage takes place in a linear sequence across time, and one of the main distinguishing features of the stages is the length of objective time each can store information. The first stage, the sensory store, holds the information for less than a second; the second stage, the short-term store, holds the information for approximately 15 seconds, unless it is rehearsed; and the final stage, the long-term store, retains information for longer periods, including the possibility of permanent storage.

It should be apparent at this juncture that cognitive psychology has come a long way from the first pioneering studies of Ebbinghaus. Nevertheless, cognitive psychology has also remained the same in many respects. One parsimonious way of summing up the similarities between past and present cognitive conceptions is to point to their common temporal metaphysic. A Newtonian approach to time (as inspired by Aristotle) remains a primary assumption. Events to be cognized take place *in* time, and cognitive processing of these events is itself subject to temporal constraints. Objective time (and space) relations still organize input from the environment. The "hardware" of the mind itself embodies the essential mechanism of Newton, with its sequential stages and uniform operation. And the "software" of the mind is the preserved, albeit encoded, associations from the past.

Rationalism Versus Empiricism

Before turning to current models of cognition, another historical influence must be explicated. Cognitive psychology also contains significant elements of philosophical *rationalism*. Rationalism and empiricism have waged a long war with one another throughout the history of recorded ideas. Part of this war concerns differing views of time (see Chapter 3). Most cognitive texts present rationalism as essentially compatible with contemporary models of cognition.[15] However, empiricism continues to dominate cognitive psychology—as we have seen—because the metaphysic of Newton continues to reign supreme. Still, cognitive psychology contains significant elements of rationalism that are the inspiration for several anomalies to be discussed later.

Modern rationalist theorizing probably begins with Immanuel Kant.[16] Kant did not view linear time as a property of the reality that formed the associations of the mind. Linear time was part of the innate *mental* organization that imposed itself *onto* a relatively unorganized reality.[17] From this framework, contiguity and repetition take on new meanings. No longer are temporal relations between events controlled by the events *in* time. The mind itself extends linear time to these events, and thus contiguity and repetition are part of the natural construction of the relations among events. This means that if associations are important to cognition, they are extended *to* rather than received *from* the world. Instead of associations forming the mind, the mind forms the associations (from birth).

The gestalt movement is in many ways an example of this Kantian line of thought. Although the gestaltists said little about time explicitly,[18] their holistic style of thinking ran counter to the reductionism of most associationists. Their famous slogan, "the whole is greater than the sum of its parts," recognizes that the relations among environmental events—temporal or spatial—cannot simply be summated associatively. There are emergent properties that cannot be understood through simple association. Moreover, these emergent wholes do not stem solely from the part-whole relations inherent in the environment. A significant portion of this holistic quality stems from relationships that the mind itself imposes, apart from previously input associations. In other words, a significant portion of the holistic structure imposed by the mind is a priori and, hence, inborn in a Kantian sense.

Perhaps the most explicit modern follower of the Kantian spirit is the linguist, Noam Chomsky. Chomsky revolutionized the study of language in the late 1950s. At the time, many considered language to be learned through an associationist process involving the contiguity and repetition of words and situations—as objectively determined. Chomsky, however, felt that many aspects of linguistic organization were innate. Although people of the world learn different languages, their inborn structures lead them to learn (and structure) the languages in basically the same way. This means that certain linguistic universals occur regardless of culture.

Chomsky also felt that this innate ability to structure the world made the learning of language more creative than associative. A child's natural ability to discern the rules of speech enabled him or her to create new sentences, rather than utter old associations. Language is learned, then, through trial extensions of self-generated structures, rather than the reception of associated structures from the environment.

The theorizing of another important historical figure—Sir Frederick Bartlett—also seems to be part of the rationalist tradition in cognitive psychology. Bartlett challenged many aspects of associationism in a provocative theory of remembering. He argued that people do not passively retrieve copies of environmental associations when remembering. He proposed instead that people *reconstruct* associations and meanings from memory fragments, much as a detective reconstructs a crime from fragments of evidence:

> Remembering is not the re-excitation of innumerable fixed, lifeless and fragmentary traces. It is an imaginative reconstruction, or construction, built out of the relation of our attitude towards a whole active mass of organised past reactions or experience, and to a little outstanding detail which commonly appears in image or in language form.[19]

Such theorizing certainly *sounds* rationalistic. Bartlett's allusion to "imaginative reconstruction" would seem to connote some *novel* contribution to the memory being recalled. Such novelty would require some influence other than the objective past (and input)— some *new* "imaginative" element being added. Cognitive texts, though, have generally interpreted Bartlett in an empiricist manner. Ashcraft, for example, views "reconstruction" as a combination of original memory fragments and already existing knowledge, both parts stemming from past input. Although the particular combination is "new" in the sense of different from the original memory encoding, its reducible "atoms" and the process of its combination are still determined by the past.[20]

Certainly no original or creative element is discussed in this regard. Nor is there postulated a cognitive capacity for formulating this type of imaginative element. All elements and processes stem from previous experiences of one sort or another.[21] Bartlett himself borrowed the idea of "schema" to explain the source of the "already existing knowledge" which added the so-called novel elements to recall. This term is sometimes associated with a Piagetian conception in which some schemata are innate and therefore rationalist.[22] However, Bartlett defined schema as "an active organisation of past reactions to past experiences."[23] This probably means that he is ultimately theorizing from an empiricist perspective, with a Newtonian framework for explanation. Similar to the historic mainstream of cognitive psychology, past input governs all cognitive systems. Even explanations that seem to emphasize present constructive or reconstructive aspects of cognition are often reducible to conventional linear theorizing.

On the other hand, there is historic precedent in the gestalt and Chomskian traditions for mental influences outside the usual linear chains of causation. Cognitivists and linguists in these traditions have access to innate and holistic approaches for establishing non-Newtonian models. Significant organizational capacities of the mind can be viewed as innate and thus ever-present, or the organizing elements of the environment and cognition can be considered to form a simultaneous whole that is "greater" than (and thus transcendent of) the sum of its associative parts (and its objective past).

Time and Contemporary Cognitive Psychology

This historic intellectual journey sets the stage for an analysis of contemporary cognitive conceptions. Multistore models of memory, at this point, have lost favor among many researchers. Conceptions in which the mind plays a more active role currently dominate the cognitive scene. Cognition is viewed as constructing, organizing, and integrating all manner of incoming information. However, the "Bartlett question" remains an issue: Are these mental activities empiricistic or rationalistic in nature? If they are empiricistic, then time and Newtonian explanation are likely to play prominent roles; past experiences ultimately are the source (and cause) of these mental activities. At the very least, some type of linear sequentiality of the cognitive process can be assumed (e.g., input $\rightarrow$ output).[24] If they are rationalistic, however, then other mental capacities (undetermined by the past) likely contribute to the cognitive or behavioral outcome.[25]

The current dissatisfaction with the multistore model has not kept it from being prominently featured in the latest cognitive texts.[26] In fact, whole chapters are often devoted to each of the three components of the model: sensory register, short-term memory, and long-term memory. As the epitome of the computer metaphor, these components incorporate all the mechanistic processes that presume a Newtonian framework for time. Each component is *reductively* independent of the other; each occurs in *linear* sequence across time; the function of each is considered invariant and *universal* regardless of the context; and the retention of information is thought to be gradual and *continuous*. Moreover, many of the important distinctions among the three components involve some quantity of *objective* time (e.g., the length of storage time and the use of repetition or rehearsal to maintain storage). Of course, the whole notion of storage takes for granted the preservation of some type of objective past.

Interestingly, the main objections to the multistore model concern these very Newtonian characteristics. In fact, these characteristics seem to be routinely challenged during transitions from one cognitive model to another. It is as if cognitive psychologists have been searching for some escape from their linear framework. All five of Newton's temporal characteristics, as manifested in the multistore model, have been challenged. Levels of processing models have challenged the reductive independence and linear sequencing of separate memory stores. Elaborative rehearsal theorists have disputed the significance of simple repetitions across time such as rehearsal. Those who consider the context of learning to be important have questioned the universality of a multistore process. And finally, some who espouse constructionism cast doubt on the storage of an objective past.

But let's not get ahead of ourselves. Let us first consider why a model with levels of processing has gained favor over the multistore model. There is certainly evidence supportive of the processing levels model, but then there is also evidence supportive of the multistore model.[27] The main advantage of the processing levels model is thought to be its theoretical parsimony.[28] Unlike a model with many storage units—each with its own properties and capacities—the levels of processing approach requires only a single memory unit, sometimes called a central processor. In this sense, retention varies with respect to the *depth* at which something is processed—the greater the depth, the greater the degree of semantic or conceptual analysis. Memory limitations do not result from limits of storage capacity but from limits of processing capacity.[29]

This processing levels model also raises questions about the rehearsal component of a multistore model. Simple repetition across time is considered necessary for an item to remain in short-term storage or be transferred to long-term storage. In contrast, different levels of processing imply that greater processing depth is the key to retention, not greater rehearsal. Craik and Watkins held that there are two types of rehearsal.[30] One simply maintains information, such as a phone number, until attention is withdrawn—the information is never really retained. The other type of rehearsal involves the processing of information at progressively greater depths. If the subject adds new associations to the items being learned, then lasting retention is achieved.

At this juncture, the levels of processing model has challenged no less than three cognitive explanations related to time—sequential staging, storage, and repetition (as rehearsal). From the perspective of the Newtonian paradigm, however, this challenge is more apparent than real. Although the model eliminates several cumbersome distinctions between stages, incoming information is still analyzed in linear stages across time.[31] Physical and sensory features are analyzed initially, with pattern recognition and abstract meaning following sequentially thereafter. The model also does not deny the storage of items in memory. In fact, stored associations and images are the determinants of the "depth" to which an item is processed.[32] And true to the thesis of repetition, this depth is not attained in one trial. Several repetitions of "elaboration" across time are necessary for lasting retention.

Of course, at this point, these models have each been revised many times, and some researchers have attempted to integrate the multistore and processing levels models.[33] Nevertheless, their fundamental assumptions of time have not changed. Past input—which is itself subject to time constraints (contiguity, repetition)—is still the driving force behind whatever variance occurs in present cognitive operations. Some observers distinguish between models that are data-driven and those that are conceptually driven.[34] Data-driven models rely exclusively on the "data" (stimuli) of the environment to effect change in cognition. Conceptually driven models, on the other hand, depend more upon concepts in cognition to drive change. Again, however, the Bartlett question is important to understand conceptually driven models: Are the cognitive concepts which "drive" the mind a product of previous experiences (e.g., associations or stimuli), or some novel mental contribution that is not totally a result of the past?

Ashcraft makes clear the fundamentally empiricistic nature of both types of cognitive models:

A data-driven mental process is one that relies almost exclusively on the "data," that is, the *stimulus information* being

presented in the environment.[35] [C]onceptually driven processing [is] where your understanding is guided by means of the
"top" level of knowledge *stored in memory*.[36]

Both types of processes employ past experiences. Data-driven
processes are determined by the "stimulus information" that occurs
in the linear sequence begun by the environment. Conceptually
driven processes stem from knowledge that is already stored in
memory. This means that the only difference between data- and
conceptually driven models is their source and temporal "distance."
The conceptually driven processes use more distant experiences
from memory storage, while the data-driven processes use more
recent experiences from the environment. Included in the concept-
driven category of models are modern renditions of Bartlett's reconstructive memory,[37] Loftus's research on eyewitness testimony,[38] semantic network[39] and integration models,[40] and script/schema
theories of memory.[41,42]

As an example, consider one of the most recent conceptually
driven theories of cognition: script theory. Its mechanistic and
empiricistic nature is clear from this description:

The overall theory behind the notion of scripts is quite straightforward: people *record in memory* a generalized representation of events they have experienced, and this representation
is invoked, i.e., retrieved, when a new experience matches an
old script.[43]

A script, then, is a representation of events from the past.
This representation is "retrieved" (a computer metaphor) when input from the immediate past (the environment) matches it. In this
sense, the representation is never *intentionally* retrieved because
this would require some capacity to do "other than" what the immediate or stored past might dictate.[44] Nor do inborn factors control retrieval variance. According to script theorists, preexisting
(past) environmental associations determine retrieval variance.[45]

Script theorists do address one of the questions raised by such
empiricistic theorizing: If preexisting knowledge is necessary to thinking and memory, how do we think and learn about situations for
which we have no preexisting knowledge? What about novel situations or items that are so unfamiliar that no relevant script, association, or schema is stored in memory? Script theorists Schank and
Abelson contend that we formulate a generalized "plan" for dealing
with "new" situations through our *past* collection of relevant sche-

mata and scripts in memory.[46] In other words, we may not have a *specific* script to deal with novel situations, but we do have more general "plans," formed from specific scripts of our past.

In answering such questions, cognitive theorists reveal how deeply they hold empiricist (and Newtonian) assumptions. Their answers have obviously not deviated from a linear emphasis upon the past. Indeed, the cognitive theorist who affirms a Newtonian framework must dispute the premise of the question—that items of learning exist for which there are no relevant past memories. Their answer must be that we *do* have *some* relevant preexisting knowledge. How else can we explain present concepts except by the past? As empiricists, they *must* question whether truly novel situations and items really exist because their existence would point to inadequacies with linear explanations of conceptually driven processing—explanations that need preexistent knowledge to account for learning.

Language and Problem Solving

Cognitive psychology contains other subjects, including language and problem-solving, that we should broadly sketch before turning to anomalies. Given the empiricistic assumptions of prominent memory models, are these assumptions important to other cognitive topics? The answer seems to be in the affirmative. Although aspects of rationalism creep into theorizing, the Newtonian paradigm seems to put any constructs that would imply a deviation from the linear and continuous past out of theoretical bounds. Nativistic propensities are clearly recognized, but they are empiricist in nature, ultimately leaving behavioral and cognitive change up to the environment (either in memory or through immediate input).

Consider the linear orientation of many contemporary psycholinguists. Linguistic inferences are often based on notions related to script theory as just described.[47] Sentences about "parties," for instance, supposedly call up from memory a "standard party script." Speech recognition is similarly viewed as a result of "concept-driven" models (e.g., spreading activation or semantic networks) in the empiricist manner defined above. Even language development is considered "continuous." Children progress from prelanguage to one-word utterances and later two-word utterances with "seamless transitions" from one developmental level to another.[48,49]

Chomsky's rationalistic influence has undoubtedly been felt. Linguistic universals are generally acknowledged, and many regard linguistic creativity as a key to language and language behavior. However, few cognitive texts ever mention Chomsky's explicitly rationalistic explanations,[50] or his own differentiation of his theorizing from comparable empiricistic formulas.[51] Likewise, his innate factors in language are rarely discussed. When they are, Chomsky is often oversimplified as a "nativist" and contrasted with an "environmentalist" linguistic position.[52] Chomsky's position (along with the environmentalist) is then cast as old-fashioned, especially in view of what is considered to be the more enlightened "interactionist" position.[53]

What is generally overlooked by cognitive theorists is that both rationalists and empiricists postulate nature/nurture interactions. From the historic debate between Locke and Kant to the more recent debate between Skinner and Chomsky, all thinkers have emphasized some interaction of our inborn nature and our environmental nurture. The key has been the *type* of nature/nurture interaction. Rationalists have viewed nature as dynamically and ever-presently active, not only in the *processing* of "information" but also in the formulation of the *content* and meaning of information. A priori cognitive organization is integral to the formation of even the most basic associations. The environment is important and influential because it provides the elements of the associations. However, the environment does not *dictate* any relationships or meanings. Relationships flow *to*, rather than *from*, the environment and our memories.

Empiricists, on the other hand, view at least *some* meanings as stemming directly from the environment, already formed through contiguous associations and the like. Our inborn nature is thus more passive. Inherited abilities do not form the basic associations between the "atoms" of our experiences (as in rationalist accounts). Nor can they *re*-form those basic atoms when reconstructing our experiences for recall. Our native abilities have more to do with the reception, storage, and retrieval of these associations *as already organized* by the environment, i.e., with the process rather than the content. This does not rule out modern models of empiricism in which previously received information influences present content. These models also require already formed content (atomic associations) which precede and thus determine how future content is affected—hence, the notion of the Lockean tabula rasa and the contemporary emphasis upon past memory stores.

The net effect is that Chomsky's rationalist position is often grossly oversimplified. Moreover, the interactionist position that is then proposed as a more enlightened approach to understanding language is really an empiricist position. The "innate equipment" proffered is the "capacity to learn and remember, with a vocal system capable of subtle muscle control, and with a perceptual mechanism that can register and retain sound."[54] This type of innate equipment merely "takes in" experience. Nothing akin to a Kantian or a Chomskian a priori cognitive organization is even implied. More importantly, it is clear in the description of this interactionist position that the environment (either through storage or stimuli) is in charge of linguistic variance. That is, if there is change in linguistic behavior, there is no ever-presently-active inborn ability that can contribute to this change. Linguistic variance is ultimately the product of the past once again.

Other cognitive topics, such as reasoning and problem-solving, follow similar lines of explanation. Although reasoning requires obvious innate abilities to do the reasoning, it is the variations of the environment that govern the variations in reasoning: "The influence of *stored information* is quite pervasive; it affects how we perform in the classic forms of reasoning as well as in less well-defined judgment and decision-making situations."[55] Explanations of problem-solving retain *some* gestalt tradition.[56] But analogous to modern psycholinguistics, those aspects of gestaltism that are uniquely rationalistic have been downplayed or overlooked. Particularly influential recently are artificial intelligence approaches to problem-solving.[57] As noted in the description of multistore cognitive models, such computer analogies fit nicely within a Newtonian framework for explanation.

This concludes our sojourn through contemporary cognitive psychology. Many aspects of contemporary cognitive models reveal Newtonian characteristics of explanation. Storage conceptions imply *continuity* in their gradual accumulation of coded associations, some of which are permanent and *universal*. This means that the past is separate from the present and essentially unchangeable. *Linear* explanations are implied in the heavy reliance upon the past and the causal sequencing of environment and cognition. Analogous to the computer, *objective* information from the environment— as formed in part through temporal contiguity and repetition— enters the mind in a linear flow of discrete bits. This flow is often then *reduced* for scientific study to independent processing stages across time.

Elements of both rationalism and empiricism are evident in many theories of memory, language, and problem-solving. However, both epistemologies have been so oversimplified and underestimated that they have lost their original philosophical distinctions. To assert an "interactionist" cognitive conception between nature and nurture is to assert very little. Both philosophies are already interactionist. The theorist must specify what *type* of interaction is being asserted. When this is specified, it is clear that empiricistic interactionism dominates, and linear theorizing is the major explanatory framework. The reason for this domination, at least in part, is the Newtonian (and Aristotelian) temporal metaphysic that cognitivists have implicitly continued to affirm.

Anomalies to the Current Paradigm

As dominating as the Newtonian paradigm has been, it has not gone unchallenged. All the mechanistic and linear aspects of this explanatory framework have been questioned. Of course, it is one thing to question these conceptions and quite another to construct alternative conceptions that avoid Newton's robust metaphysic. Still, anomalous frameworks for cognition have been formulated. In fact, there appear to be three major sources of anomalies, all outside mainstream cognitive psychology. Each stems from more rationalist lines of thought and overlap somewhat for this reason.

A purely rationalistic approach to constructivism is first discussed. Although this more Piagetian line of thinking is elaborated somewhat in Chapter 2, its cognitive and temporal implications require discussion here. Second, many humanists have reacted unfavorably to mechanistic conceptions. The cognitive implications of humanism, especially those concerning time, are next explored through the work of Joseph Rychlak. Finally, alternative cognitive conceptions have originated from those disputing the advances of artificial intelligence (AI). AI is the other side of the cognitive coin: If the human mind is essentially a computer, then the computer is (or has the potential to be) an "artificial intelligence." A critique of AI, as exemplified in the work of Hubert Dreyfus, is therefore briefly sketched.

Rationalistic Constructivism

Let us begin with constructivism. By its very nature, constructivism implies an interaction between the entity doing the constructing and the thing being constructed. Recall, however, that there are different types of interactions. Mainstream cognitive constructivists, in the tradition of Bartlett, generally endorse an empiricistic interactionism in which linear factors, such as past inputs and stimuli, ultimately govern construction. There is, however, a *rationalistic interactionism* that implies a rationalistic constructivism. Theorists in this tradition are less dependent upon Newtonian characteristics of explanation. Piaget is perhaps most noted for his formulation of this tradition, though more recent theorists have extended his thinking into more "radical" constructivist realms.[58]

Above all else, Piaget is a champion of the *interaction* between persons and their environments. But he specifically rejected the interactionism found in computer analogies and empiricism.[59] He felt that this mode of explanation is ultimately not interactional, giving the environment too much influence. He also did not believe in a "one-sided" subjectivism of the mind, as Bandura would say (see Chapter 3). Piaget asserted a constant interaction between the person and environment *without* the temporal sequencing inherent in Newtonian approaches. As he put it: "Knowledge does not *begin* in the I, and it does not *begin* in the [environmental] object; it begins in the interactions . . . there [being] a reciprocal and *simultaneous* construction of the subject on the one hand and the object on the other."[60] His notion of "simultaneous" is crucial here, because the influences of the environment and the person are concurrent, existing in the same moment. Neither factor precedes the other in time. Therefore, neither factor can take precedence over the other—at any point in time—and a rationalistic interaction is possible.

Piaget realized that such theorizing is only possible if both the person and the environment can be influential *from the start*. No empiricistic tabula rasa can be permitted, because this implies that the person is not influential from the start. "Conceptually driven" models do not help here either, because the concepts that drive processing are themselves driven by the environment earlier in time (see above discussion). In contrast, Piaget realized that some *innate* ability to structure knowledge is needed. He was careful not to postulate the Platonic notion that knowledge itself was inborn. Knowledge is interactional in his estimation—a product of constructions. Nonetheless, this implies that the mind has biological structuring abilities that are at least *initially* independent of the structures of the environment (as the environment is, in some sense, independent of the mind). The sucking response is one such example. The child sucks at the mother's breast instinctively, and then assimilates this action to other related actions, such as sucking the thumb and ultimately "sucking" the world.[61,62]

This theorizing allows Piaget to avoid more than just the chronological sequencing of empiricistic interactionism. His notion of the simultaneity of mind and environment also permits the two entities to form a gestalt. Without simultaneity, the two entities can never be present together to form a single whole. They are always separated in time and thus independent of one another.[63] Simultaneity, on the other hand, allows for the possibility of a new pattern of relationship (emergence) which is not merely the sum of its elements. Indeed, the elements in such emergent wholes are not the important factors. As Piaget noted, it is rather "the relations among elements that count."[64] In this sense, concurrent relations do not have to be continuous with their past. They can emerge into qualitatively different patterns that are not possible among elements separated in time. Therefore, empiricistic interactions can never effect qualitative differences because their elements are never together in time (e.g., Bandura's reciprocal determinism, see Chapter 3).

Piaget reflects another divergence from Newtonian temporal conceptions in his conception of change. Neither innate nor acquired structures are universal and unchanging across time, according to Piaget. Recall that in empiricistic interactions the inborn capacities and the information stored in long-term memory are relatively permanent and unchanging. The genetic "hardware" is relatively fixed across time, and the informational "software"—at least in an adult "program"—is also fairly well established. More information is constantly being added, but even it is interpreted in light of the *existing* hardware/software interaction, as mentioned in the previous section.

Piaget, on the other hand, asserted the existence of active innate capacities. These capacities are dynamic, changeable, and constantly imparting their own influences to experience:

> Whereas other animals cannot alter themselves except by changing their species, man can transform himself by transforming the world and can structure himself by constructing structures; and these structures are his own, for they are not eternally predestined either from within or from without.[65]

Piaget is pointing here to holistic and qualitative changes in cognition. Lower animals are relegated to passive genetic predispositions that fundamentally cannot be changed. Humans, however, have capacities to alter their constructions in ways that the environment does not determine. These changes are not simply changes in a developmental stage. It is the nature of cognition to construct its perception and its memory *constantly*. Memory, then, is subject to the vicissitudes of momentary cognitive constructions. That is, memory is in a continual state of moment-by-moment *reconstruction*. The apparent universality and objectivity of experience—either as a set of stimuli or as a permanent memory store—is an illusion of the adult construction of an objectified reality.

More recent rationalistic constructivists have made this clear. Melkman, for example, describes this in her book, *The Construction of Objectivity*. She notes that the temporal sequencing of empiricism leads to a subject/object dichotomy. To assert that one entity comes *before* the other is to assert the independent existence of each. The fact that the object of the environment is the entity "before" means that this object must be received by the mind and then internally represented as knowledge. As Melkman observes, however, "knowledge must be described in the context of its formation and from the perspective of the *momentary* subject toward its object."[66] This means that the subject and object are never separated, but must be understood as one entity in the same "moment," rather than two entities in a linear sequence.

The "objects" of an adult environment are mentally constructed at each moment of their perception. The structure of the environment is never "out there." It is constantly being reconstituted by cognition, and is therefore liable to change as the structures of cognition themselves shift. This is also true for the "objects" of memory. Without the objectification of the environment, no replicas of the environment can be stored in memory. "Replicas" are also constructed (or reconstructed). Unlike empiricistic interactionism, prior storage of information (supposedly unconstructed at some earlier time) does not determine reconstruction. Reconstruction is instead guided by *present* constructions, themselves originally (and continually) constructed by active, innate structures and thus not solely governed by the environment.[67]

How does this square with the data on the reconstruction of memory? Most of the data on this issue have been collected and interpreted from the viewpoint of an empiricistic constructivist. Can a rationalistic constructivist subsume these data? Most, if not all, of these data can be "reconstructed" to support a rationalist perspective. Research on mood congruence, for example, can be reinterpreted as evidence for this perspective.[68] Mood congruence refers to the common finding that people are more likely to remember information that is congruent with their mood state.[69] A happy person is more likely to remember happy incidents, and conversely, a sad person is more likely to remember sad incidents. The past is, in this sense, reconstructed in light of a *present* mood.

Mainstream cognitive explanations of these findings are clearly empiricist and associationist in nature. State-dependent learning, for example, hypothesizes that many items of memory are associated (or encoded) with a certain mood state at the time of their learning. This means that the mood association serves as an effective cue for future recall.[70] If, for instance, the learning of a story passage is contiguous with a sad mood, then subsequent recall is facilitated if one is again in a sad mood. True to a Newtonian framework, the objective properties of time figure prominently in this explanation: (1) the mood and item are considered to form a temporally contiguous association, and (2) this association is thought to be preserved objectively in memory, ultimately governing its recall.

A more rationalistic approach, however, postulates that contiguous associations from previous "objects" of the environment are not preserved in this manner. They are reconstructed in light of the *present context*. That is, even the so-called atoms of the empiricist approach (past associations) are subject to present constructive processes. This would imply that present mood-states can themselves contribute to the creation of new associations (or meanings). Rather than the previous mood association being stored as an aspect of a memory, the present mood itself could construct a new meaning of the past memory.

For example, a currently happy person could reconstruct an incident learned in sadness with a positive emotional tone. What might have been a devastating childhood trauma could be reinterpreted as a good lesson learned (from a more positive emotional perspective). A totally new "association" would thus be born. Depending upon the mood, the same instance could be *re*interpreted as negative in emotional tone. The point is that the present and past are conjoined, with the present playing an *active* role in the process of reconstruction. This type of explanation would account for most of the findings gathered from a more empiricistic perspective,[71] as well as many findings on mood congruency that indicate its independence from linear constructs such as past associations.[72]

Humanistic Learning Theory

Humanists have varied greatly in their theoretical positions, so any attempt at general characterization is hazardous. Nevertheless, one theorist and researcher, Joseph Rychlak, has proposed a humanistic epistemology that he *explicitly* contends is an alternative to the usual Newtonian framework.[73] He first notes the significance of absolute time to modern cognitivists. "We cannot overestimate the importance of the time dimension to explanations of behavior proffered by the learning theorist (which includes not only S-R [Stimulus-Response] but cybernetic and information-processing accounts as well)."[74] He then argues that a humanist would need to establish a framework for learning that is outside this time dimension. Humanists have traditionally questioned the notion of objectified entities, such as the environment, determining behavior and cognition from the past. Their focus has been more present-oriented, both in theory and therapy.

Why? According to Rychlak, humanists view humans from a teleological perspective. The "cause" of the person's thoughts and behaviors for the humanist is the "telos" or goal "for the sake of which" the person behaves. This type of cause (sometimes termed "final cause") does not stem from the immediate past, as in the sense of a stimulus, or the distant past, as in the sense of stored information or associations. The final cause is a *present phenomenon only.*[75] Goals and intentions may be formulated for the sake of a "future" image, but the image itself is a present one affecting present behavior and thoughts and is subject to change in the next instant.[76] The person, then, is the moment by moment determiner of the intention "for the sake of which" he or she behaves.[77] Intentions may look continuous, but this is because the same intentions are being affirmed across time.

Rychlak acknowledges that many cognitivists frequently theorize in telic terms and constructs (e.g., reasons, intentions, and expectations). He notes, however, that the mere use of such terms does not make theories telic, nor does their use imply that such theories account for the telic concerns of humanists. Cognitivists typically consider telic constructs to be themselves determined by the past in some manner. Goals, intentions, and expectations are thought to be caused by previous experiences with the environment. Telic constructs are causes only because they are links in the causal chain from the past. Humanists, however, regard telic constructs as *original* causal contributions to cognition. They are not mere by-products or the end-products of linear cause-and-effect chains; they are the initiators of cognitive patterns.

How would memory operate without linear time? Rychlak holds that learning and memory occur through logical rather than chronological relationships. He proposes a "Logical Learning Theory" that relies on the "timeless" aspects of logic (e.g., tautology, predication, and dialectic) to explain cognition. Logic has long been held to operate without temporal duration or sequence (see Chapter 4). For instance, the relation between two premises of a syllogism (the logical conclusion) occurs simultaneously with their presentation. The premises, "All horses are animals" and "Flicka is a horse," can be represented by appropriately labeled concentric circles (cf. Figure 4.1). When the premises are drawn in this manner, the conclusion, "Flicka is an animal," is *already* implicit, or as Rychlak puts it, "tautological" with the two premises.[78] Temporal duration and sequential processing are not required to learn the conclusion. Simply realizing the relation between the premises is sufficient. Newtonian machines, on the other hand, necessitate processing across time to arrive at conclusions.[79]

According to Rychlak, the teleologist views the learner's cognitive organization and the organization of the information to be learned as being analogous to syllogistic premises (or parts of a whole). Similar to Piaget, this relation takes place concurrently; the environment is not chronologically first. If anything, the mind is *logically* precedent because it formulates the intention "for the sake of which" behavior is carried out. The intentional organization of cognition is wedded to the structure of the environment (like two premises of a syllogism), and this wedding determines what is learned (the logical conclusion).[80] Aspects of the environment that are relevant (or tautological) to the person's internal cognitive organization are those that are learned most readily. That is, meanings (information) related to the person's goals are the most meaningful. Car commercials, for instance, suddenly become meaningful when one decides it is time for a new automobile. The reason that goal-related meanings are learned the most readily is that these meanings are *already implicit* in the cognitive organization of the learner.

Learning, then, is not some gradual (and continuous) accumulation of previously unknown "facts" and associations. *Learning is an elaboration of what one already knows.* To a linear theorist, of course, what one "already knows" must entail some past input or information. To a humanist, such as Rychlak, however, the "already known" can be inborn or even cognitively invented in the present (as a telic image of the "future"). The primary point here is that the "already known" is *implicit* in (and tautological with) the cognitive organization of the learner. No temporal sequence or primacy of the past is required. Just as the conclusion of the syllogism about horses is an elaboration (or an application) of the major premise—all horses are animals—so the learning of the mind is an application of its own intentional premises.

One does not learn to ride a bike, for example, by committing the facts of a bicycle to memory. Knowing the mechanics of a bike or the physiology of our leg muscles does little to help us actually ride. One learns by taking the *already known* structures of balance and motion, and applying them to the unknown (but analogous to the known) realm of bicycle riding. Nothing "new" in the sense of input is learned, because bicycle riding—and for that matter, all learning—is already implicit in cognitive schema which in this case is inborn. Although the full pattern of riding typically takes several trials (and physical development), Rychlak would say that some aspects (patterns within the patterns) are learned at each trial in which mental and environmental factors are logically conjoined. Repetition is therefore not integral to the learning per se; repetition merely permits the elaboration of premises to occur.[81]

Rychlak has conducted over a hundred studies that he feels support these assertions.[82] They encompass the more strictly controlled laboratory experiments as well as the more applied investigations. In many of the studies, he and his students have attempted to control factors considered important to empiricist explanations. For example, they have attempted to control and manipulate prevalent associationist factors in meaningfulness: frequency of exposure, familiarity, number of associations, rehearsal, practice. Rychlak and his associates have repeatedly found that such factors do not account for variance in meaningfulness and memory. In other words, objective past experiences are not as significant as most cognitivists have assumed. Many cognitive capacities (e.g., affective assessment and predication) do not rely on such influences, according to Rychlak.

One of the capacities to which Rychlak alludes is free will. Free will or the "ability to do other than" transcends the objective past, by definition.[83] This is not to negate the importance of "past information" in the formulation of one's will, but the past cannot *determine* free will or it is not "free." Free will also implies the possibility of dramatic discontinuities in behavior and cognition. A person can choose (doing "other than") a completely different course of action than that consistent with past experiences. It is not by coincidence that only rationalists like Rychlak are willing to entertain the possibility of free will. Most other cognitive researchers are engulfed in Newtonian conceptions that make determinism of the past a foregone conclusion. This, in turn, prevents formulation

of transcendent capacities or cognitive abilities that are not determined by the objective past.

The Black Knight of Artificial Intelligence

Researchers in artificial intelligence are working on the flipside of
the Newtonian paradigm in cognitive psychology. Because
cognitivists have contended that the human mind is essentially a
form of computer,[84] many AI researchers have contended that the
computer can become an intelligence essentially analogous to a
human intelligence. This contention has led to a flurry of research
on computers. Visions of robots, androids, electronic chess masters,
and computerized teachers and therapists have excited scientists.
Many have considered the computer age a technological revolution,
and predicted vast changes and major breakthroughs in understanding.[85]

The "Black Knight of AI" Hubert Dreyfus,[86] however, expresses
no small amount of skepticism about such predictions.[87] Indeed, he
has chronicled the progress of AI in fulfilling its claims and finds
that it comes up dramatically short. He contends that this curious
lack of progress is not simply the result of overoptimism regarding
the speed of these breakthroughs. It is rather a misconception about
the way in which the human mind operates. Although Dreyfus does
not specifically mention the issue of time, he proposes a distinctly
"nonmechanistic model of human skill."[88] He specifically challenges
two linear characteristics of explanation that have long been assumed to be integral to models of human cognition: reductionism
and universality.

Dreyfus notes that all information processing models assume
that data from the environment are received "in terms of atomic
elements logically independent of one another."[89] Regardless of the
processor, human or computer, the input is portrayed as a flow of
discrete *bits* (or *b*inary dig*its*) across time—individual pieces of
information that are processed one by one. What Dreyfus does not
mention is the Newtonian basis of this portrayal. The assumption
of linear time implies that all processes occur piece by piece across
time. As we saw with developmental processes in Chapter 2, this
leads to a form of *reductionism* that considers each piece separable
from (and sometimes causal to) other pieces.

The unfortunate consequence of this view is that the whole of
processes is often not studied, or studied with immense difficulty.
In the investigation of development, for instance, researchers have

found it difficult to discover the relations *among* individual age groups. Considerable information has been gathered about specific stages or ages (cross-sectional research), but this type of information does not tell researchers how a child moves from one developmental stage to the next (see Chapter 2). The problem is the same for understanding any process across time. This includes understanding information that supposedly occurs bit by bit across time. Because the process does not occur *at the same time*, the whole of the process (or its meaning) must be reconstructed from recordings (or symbolic representations) of the parts across time.

This type of reductionism is a major problem for artificial intelligence, according to Dreyfus. As Anthony Oettinger puts it, "the burden of artificial intelligence is indeed its apparent need to proceed—in futility—from the atom to the whole."[90] In the same way that developmentalists cannot understand the whole of development from independent age groups, machines cannot understand the meaning of information from independent data bits.[91] Computers cannot reconstruct the relations between the parts (the whole) when the parts are presented as "logically independent elements." Information about their "betweenness" is missing. Because the whole is different from the sum of the parts, holistic qualities can only exist when the parts *are* a whole—i.e., when the parts exist *concurrently*. When each part is presented *separately*—as in a linear flow of information—the singular nature of their holistic relation is lost. The meaning of the pattern must be reconstructed from an assemblage of bits that could have several possible interpretations.

From Dreyfus's perspective, however, humans do not have this problem. The reason is that humans can naturally perceive all parts of the whole *simultaneously*, rather than each part of the whole in linear sequence. Even if humans attend primarily to component parts, the holistic quality of each part is usually evident. The parts are seldom seen as "atoms"—independent of one another. In recognizing a melody, for instance, the notes get their values by being perceived as part of the melody, *not* by the melody's being recognized in terms of independently identified notes. Likewise, the total meaning of a sentence determines the value (or meaning) to be assigned to the individual words. According to Dreyfus, though, computers lack the ability to perceive relationships and thus cannot discern complex meanings and patterns.

AI researchers have responded to this problem by claiming that holistic qualities can be formalized into a set of heuristic rules which are input into the computer to direct the relationships among the individual bits. This formalization is the other mechanistic and linear characteristic challenged by Dreyfus—*universality*. Recall that universality is the Platonic notion that some unchangeable and timeless essence stands behind the constantly changing world of our everyday experiences. Absolute time is integral to this notion because its uniformity conveys the sense that natural processes are essentially the same, regardless of their temporal context. If time were not uniform, then the essence of natural processes would be unstable and, indeed, specific to the temporal context in which they manifested themselves. Knowledge that transcends the specific context would be impossible.

Dreyfus demonstrates that AI researchers make universal assumptions about knowledge. They assume that the heuristic rules which direct the organization of informational bits are not specific to the particular context. AI researchers assume that it is just a matter of discovering the general rules of these contexts, so they can be formalized and input into the computer (or person).[92] Dreyfus, however, disagrees with such assumptions. He questions whether the holistic qualities that are simultaneously perceived by humans can be formalized into general rules.[93] He questions this on several grounds that space constraints prohibit us from reviewing comprehensively. Nonetheless, let us attempt to describe one of these so that the reader may have a sense of Dreyfus's criticism of Newtonian universality.

Fundamentally, Dreyfus contends that holistic "information" regarding the temporal context of perception is not formalizable or determinate. First, there is the basic figure-ground relationship necessary for any perception. Whatever is prominent in our experience (the figure) appears simultaneously against a background or surrounding context (the ground). As gestaltists have long recognized, the figure has many determinate features. The figure is a unified, bounded entity which has features that can be described and presumably formalized into a set of universal rules. The ground, on the other hand, is essentially indeterminate. The ground is necessarily fuzzy and vague by its very nature; it is not bounded or unified. Indeed, we do not even attend to it. As Dreyfus puts it, the ground can only be characterized as "that-which-is-not-the-figure."[94]

This means that the ground *cannot* be formalized. The surrounding context of the figure cannot even be positively described, let alone specified in the precision necessary for a linear processing machine. If it is attended to or examined, then *it* becomes the figure and the ground is lost once again. Some have advocated not worrying about the ground, and many AI researchers have proceeded as if information about the figure were the only issue. The figure, however, derives much of its own meaning from the ground of its surrounding context. That is, even though the ground is indeterminate and, in some sense, not "seen," it nevertheless affects what *is* seen. As Merleau-Ponty expresses it, "The perceptual 'something' is always in the middle of something else; it always forms part of a 'field.'"[95] The ground is therefore integral to the figure. Without determinate and specifiable knowledge of the ground (which is indeterminate by definition), the meaning of the figure is lost as well.

Unfortunately, information processing machines (and their models) operate only on completely determinate and linear data.[96] Such machines must process even pictorial information in small bits and cannot therefore "see" crucial holistic qualities, such as the meaning of a figure. This is actually the reverse of humans; humans see the whole and then the details within the whole's context. The binary logic of computers, however, makes their processing mode either on or off. Given any aspect of a perceptual scene, the machine either picks up the first piece of information it encounters, or it does not. All ensuing information about the object (as it is searched visually) is added to the discrete bits already stored in memory. The upshot is that artificial intelligence cannot, in principle, operate like human intelligence. Information processing requires a reductionism of the information that loses the holistic qualities of the parts, and many of those holistic qualities, such as the ground of perceptions, cannot be formalized for linear input.

What does all this mean regarding Dreyfus's alternative to linear cognitive models? Dreyfus is offering psychology a means of understanding cognition without the normal Newtonian assumptions. Information is not handled mentally in a temporal sequence, either in its reception or in its processing. Cognition is more of a gestalt "processor." Similar to Piaget and Rychlak, the elements of the information must be "co-present" to be really understood.[97] Dreyfus contends that the role of our bodies in coordinating the simultaneous aspects of perception is grossly underestimated.[98] Moreover, not all knowledge is objective or universal. Crucial aspects of our understanding are not programmable, and so are not being "input" or "stored" in any meaningful sense. Crucial aspects are endemic to the situation itself, making understanding temporally *situated* in, and not transcendent of, its unique context.

Conclusion

Several conclusions about the temporal framework of cognitive psychology can be drawn. First, the significance of a Newtonian metaphysic is clear in the early history of the discipline when objective temporal properties were thought to control associations. Despite recent movements away from original associationistic conceptions, objective temporal properties, such as contiguity and repetition, remain integral to contemporary cognitive conceptions. Second, empiricism still dominates mainstream cognitive psychology. Greater stress has been placed upon the activity and interactionism of the mind, but this activity is ultimately controlled by the environment—either through storage or immediate input—and the particular interactionism championed is empiricistic in nature.

From this perspective, cognitive psychology is not really revolutionary. It has liberalized behavioristic method and theory to include the "software" of the mind, but it still relies exclusively upon mechanistic metaphors. True to the popularizer of machine metaphors—Newton—it preserves every characteristic of his temporal framework for explanation. Time is viewed as *objective*, allowing for the formation of contiguous associations as well as methods that rely upon learning rates and reaction times. Mind is also considered to operate in some sort of *linear* sequence across time. Although the more obvious purveyors of such sequences (e.g., multistore systems) have found disfavor, all the more recent models evidence temporal linearity beginning with the environment. This makes the primacy of the past fundamental to any theorizing.

The other three characteristics of a Newtonian framework—continuity, universality, and reductionism—are less explicit in cognitive theorizing but no less pervasive. Often it is by contrast with alternative characteristics that they become evident. Dreyfus contends, for example, that many aspects of the process and content of the mind are indeterminate and unique to temporal contexts. This helps reveal the prevalent contrasting assumption that these aspects are *universal* and unchanging across time. Moreover, cognitive explanations almost never entail discontinuities in learning and memory. The mind is thought to be a *continuous* storer of facts and rules; information in general is gradually accumulated over time. Holism too is rarely integrated into cognitive models. This is because such models rely upon mechanistic "processing" that *reduces* software and hardware to logically and temporally independent entities across time. Without their concurrent relations, these entities cannot possess holistic qualities.

Even so-called active features of the mind are portrayed in linear terms. Many cognitivists, for instance, discuss the constructive and integrative nature of the human mind. The mind is described as constructing its perceptions from the sense data of the environment, and reconstructing its remembrances from the memory fragments of previous input. However, the Bartlett question is pivotal in such theorizing: What causes or controls these constructive and reconstructive processes? The repeated answer to this question reveals the fundamental empiricism and linearity of cognitive psychology: previously stored information. Even models that are regarded as "conceptually driven" ultimately have the same originating and controlling factors as the "data-driven" variety—i.e., past input.

Anomalous models—though well outside mainstream cognitive psychology—demonstrate how theorists have attempted to escape this dependence upon past input. Piaget describes a rationalistic interactionism which questions the objectivity of our experience, and gives the person the power to transcend the objective past. Subject and object are not separated in time, and thus are simultaneously and momentarily whole. The humanist, Rychlak, explicitly advocates a learning theory devoid of absolute time. No temporal sequence or duration is involved in his model, and learning is not a continuous accumulation of independent facts. Learning is an extension of already implicit meanings, controlled to some degree by a free will that escapes the determinism of past input. Dreyfus does not challenge time as directly as Rychlak, but he does question the appropriateness of mechanistic models of human cognition. His alternative approach posits relational properties of the information and the "processor" that cannot be mechanistically specified or previously input.

Chapter 6

INDIVIDUAL THERAPY

When surveying the various modes of psychotherapy, it is striking how much individual therapy dominates the clinical scene. In fact, some training programs in clinical or counseling psychology focus almost exclusively on individual psychotherapy. Part of this dominance is due to the most frequently cited characteristic of our society—"individualism."[1] Individual therapy requires only one person (along with the psychotherapist). Participants in group and family therapies must work in conjunction with other nonprofessionals (see Chapters 7 and 8), whereas individual therapy clients do not have to suffer the complication and embarrassment of their revelations to other therapeutic participants. Clients in individual therapy can more or less "make it on their own."

A major reason for the dominance of individualism is a recent historical change in the way that people are understood.[2] The historic rise of linear time is one of the main factors in this change (see Chapters 1 and 2). Before the appearance of absolute linear assumptions, people were understood more in relation to each other in their present context. Society saw itself more as a single (and present) corporate body than as a collection of individuals with unique pasts. Any understanding or treatment of psychological maladies concentrated on present circumstances and relationships. After the rise of linear time, however, people were understood more in relation to themselves as individuals in the past. The emphasis of Newtonian science upon cause-and-effect chains and continuous/universal stabilities across time placed emphasis upon each person's unique past for complete explanation in the present.

In this sense, all orientations that emphasize individual treatment may be Newtonian to some degree. This is not to say that therapists cannot treat individuals while taking into account their

139

holistic (and present) context (see Chapter 8). Nevertheless, most therapists emphasize the individual's unique immediate or distant past in their understanding, and thus reduce psychological problems to the individual who supposedly carries this past from place to place. Freud's emphasis on past experiences is partly responsible for this. As Reeves puts it, "what Freud did was legitimize and, eventually, institutionalize an emphasis on the individual and the self."[3] Behavioral therapists have also reinforced this legitimization, despite their recognition of environmental forces. The individual remains the focus because the individual is the storehouse of past environments through reinforcement history or cognitive programming.

As strong as this individualism is, it has not prevented Newtonian anomalies from arising in the field of individual therapy. It is rare that the explanatory framework of Newton is entirely rejected, but several approaches appear to challenge substantial portions of his temporal assumptions. Prominent among these are so-called third-force approaches. As early as Carl Rogers, third-force theorists have tended to focus upon the present rather than the past in understanding patients. Although discussion of the past is never ruled out, the past is rarely viewed as the determiner of the present. Indeed, the present or future, as perceived by clients, is seen to be more influential. Clients are routinely considered to be discontinuous with their pasts, and so therapists have assumed that few universals across time (e.g., laws) apply.

This chapter begins with a conceptual history of individual therapy. It basically picks up where historical lessons in the personality and cognitive chapters (3 and 5) leave off—subjectivist and objectivist views of linear time. These views are highlighted in the historical precedents they set for individual therapy. Then, the current implications of these precedents are described. Modern conceptions of psychoanalytic and behavioral therapies are shown to follow essentially objectivist assumptions. Even the more theoretically liberal notions, such as the cognitive therapies, are revealed to assume fundamentally objectivist principles concerning time. The only individual therapies that seriously jeopardize the dominance of objectivist temporal assumptions are some third-force approaches, such as existentialism, which few therapists practice and even fewer practice with consistent subjectivist assumptions.[4]

The History of Individual Therapy and Time

In many ways, the history of individual therapy is merely the applied side of the histories of personality and cognitive psychology (see Chapter 3 and 5).[5] Similar temporal distinctions apply. The primary one is the distinction between objectivist and subjectivist views of time. The objectivist considers linear time to be absolute in the epistemological tradition of John Locke, the self-admitted "underlaborer" of Newton. This means that linear time is objectively existent—independent of our perception and our conception. Indeed, all the events of our perceptual and conceptual apparatus are themselves born by the medium of time. The mind mechanistically "processes" information *across* time. Events become associated by the mind because of their contiguity *in* time. Certain associations are particularly strong and meaningful because of their repeated exposure *through* time. And in general, the mind is a storage and retrieval center for our personal pasts.

The subjectivist, on the other hand, considers linear time to be more a product of the mind than the mind a product of linear time. Unlike Locke, Immanuel Kant did not view linear time as existing independently of the person's conceptual apparatus. Linear relations are part of the organization used *by* the conceptual apparatus. To consider the mind to be subject to linear constraints is to forget that the mind *generates* the linear constraints in the first place. The so-called medium of time does not exist. The primacy of the past, as well as related notions like linear causality and temporal contiguity, are extended *to*, rather than received *from*, world events. Our minds and personalities are not products of the storage and retrieval of either our distant or our immediate pasts (e.g., stimuli). Instead, sensations and memories are conceptualized and placed within a linear context *as* they are known. In this sense, the present and future influence the mind as much as the past.

Interestingly, it was the admixture of these two great traditions of time that began modern individual therapy in the work of Sigmund Freud. Many commentators have cast Freud as a thoroughgoing objectivist, yet significant elements of subjectivist assumptions pervade, if not dominate, his theorizing (see Chapter 3). His objectivist side is perhaps the more familiar. He clearly valued the past in understanding the present problems of a patient. He also considered the sequential development of the person to be paramount in understanding his or her behaviors and emotions. Nonetheless, Freud's conceptions of the past and development are not always linear and mechanistic. The past is more subjective than objective, and more dynamic and changeable than static and stable. For Freud, events of the past are first interpreted by the individual at the time of their occurrence. And later, these interpretations are themselves continually *re*interpreted, as the meaning of the person's "past" is constructed in light of the meaning of the person's "present."

How then does this mixture of subjectivist and objectivist assumptions affect Freud's therapy? Again, the objectivist elements are the more familiar because our Newtonian culture has given the most attention to these. First, many of Freud's techniques seem to be for the purpose of temporal excavation, i.e., digging into the past. Free association, dream interpretation, and the facilitation of transference all serve as instruments of temporal discovery. Second, psychoanalytic interpretation of a patient's behaviors requires extensive knowledge of universal developmental stages and fixations. This means that Freud felt he had access to timeless universals of personality which allowed him to assume the expert role in any therapeutic encounter. This expert role is a clear Newtonian and objectivist implication. Given the uniformity and predictability of absolute time, universal laws of therapy can be formulated and employed for every client.

Freud's subjectivist elements are the most evident in his theory of cure. In fact, without some element of temporal subjectivism, Freud's approach to treatment would have been ineffective. Patients determined by inputs from their past can only be *re*determined through a similar process. If patients were determined by some sort of past "programming," then the logical treatment to counteract this would be some sort of "*re*programming," as in some behavioral treatments. The problem with this rendering of Freud is that he never advocated such a direct and active approach for the therapist. On the contrary, orthodox analysts are the epitome of nondirective and nonactive. They are more interested in their patients discovering the meanings of their past (through insight) than in *directly* changing any mental structure. Indeed, according to Freud, therapists are not capable of directly manipulating the mental structures of their patients.[6] Patients can resist the actions of their therapists. In his view, the mind does not have a linear connection to its environment.

The mind also has no linear connection to its past. It is the *meaning* of the past that results in the many intrapsychic difficulties faced by so many patients. In subjectivist parlance, the goal of Freud's treatment is the reconstrual of these meanings (something not considered possible by purely objectivist standards). Once the meaning of the past had been brought into the conscious realm, it is then possible to recast this meaning. With a transference reaction to the therapist, for instance, the patient can see the meanings of his or her past displayed behaviorally. This leads to a new awareness (insight) which can occur *discontinuously*. That is, the move from the conscious to the unconscious realm can be sudden and instantaneous, with an entire gestalt of meanings abruptly recognized. Conscious recognition of the past provides conscious control, and alternative meanings can be worked through with the therapist. The patient is then considered to be capable of acting for the sake of a new set of meanings.

This "acting for the sake of new" meanings refers to another subjectivist feature of Freudian therapy—teleology. Teleology is the future-oriented property of many Freudian meanings.[7] Although Freud valued the meanings of the past, he did not overlook the meanings of the future.[8] Moreover, reconstrual of the past does not mean merely a different understanding of the past. It means a new set of goals and intentions for the present and future. Freed of the binding influences of past meanings, persons can more productively act for the sake of their future. As he put it in his discussion of parapraxes (i.e., Freudian slips):

> All those of us who can look back on a comparatively long experience of life will probably admit that we should have spared ourselves many disappointments and painful surprises if we had found the courage and determination to interpret small parapraxes experienced in our human contacts as auguries [signs, clues] and to make use of them as *indications of intentions* that were still concealed.[9]

Freud is contending here that our unconscious intentions and wishes for the future are important. Indeed, he seems to imply that our lives could proceed down alternate routes (deviating from the linear past) if we could become conscious of these goals and objectives for the future. In fact, the commentator Joseph Rychlak asserts that teleological (or final causal) determinism is the more appropriate way of understanding Freud's *psychic determinism*.[10] The psyche, in this sense, is more determined by its intentions and wishes than its past experiences.[11] Instead of events from the past determining parapraxes, unconscious wishes and intentions for the future (themselves undetermined by the past) cause these Freudian slips. In this manner, the future is as much a determinant of the present as the past. Such telic themes are given even more prominence in the theorizing of Freud's psychoanalytic peers, Jung and Adler, as discussed below (see also Chapter 3).

This brings us to a fascinating aspect of the legacy of Freud and the history of individual therapy—namely, its intolerance for his mixed model of time. Therapists have seemed to emphasize either Freud's objectivist views or his subjectivist views of linear time but *not both*. The objectivist side of Freudian theorizing—his emphasis upon the past, linear sequencing, and universalized aspects of personality—was amplified in the explanations and techniques of Dollard and Miller and eventually the behaviorists. The subjectivist side of Freudian theory—reconstrual of the past, past as context and not determinant, "future" determining present— was emphasized in the theories and therapies of Jung, Adler, and ultimately the phenomenologists and existentialists.

Dollard and Miller, as noted in Chapter 3, are transitional theorists. They consciously attempted to translate all of Freud's subjectivist constructs into objectivist constructs in the name of Newtonian science. They succeeded to a large extent. Psychotherapy ceased to be a discovery and reconstrual of past meanings, and became, as Rychlak puts it, "a matter of unlearning old habits and eventually learning new habits to take their place."[12] For Dollard and Miller, patients are "programmed" incorrectly by their past. The therapist's task is to use reinforcement techniques to do the reprogramming required. The role of the therapist moved from a listener and interpreter of the patient's meanings to a teacher and coach of the proper mental and emotional habits. "In the same way and by the same principles that bad tennis habits can be corrected by a good coach," declare Dollard and Miller, "so bad mental and emotional habits can be corrected by a psychotherapist."[13] This approach is the forerunner of modern behavioral therapies.

Dollard and Miller, however, did not concentrate solely upon behavioral habits. Their emphasis upon "mental and emotional habits" points to another important objectivist influence in modern individual therapy—*mediational theorizing* (see Chapter 5). Dollard and Miller were among the first therapists to view cognition as a mediator between the environment coming "in" and the response (emotion or behavior) going "out." They did not deny the significance of *non*mental habits built up across time, such as conditioned behaviors and direct stimulus-response (S-R) associations. Nevertheless, they felt that the job of the therapist is to help the patient mentally mediate these habits, so they can be brought under cognitive (conscious) control. Providing a label for an S-R association, for example, helps this to occur. In a sense, it allows for more "insight" by permitting finer discriminations between stimuli.

Despite the Freudian and cognitive terminology (e.g., insight or conscious control), objectivist assumptions remain in force. In fact, cognitive mediation for Dollard and Miller is itself a mechanism for preserving objective temporal properties. The mind accumulates previous stimulus-response associations (cue-producing responses) so that new stimuli can be responded to appropriately. The mediator does not stray from its past to contribute anything new. Indeed, the mediator is merely one link in the causal chain begun by the environment. This leaves it under the immediate control of the previous "cause" as stimulus and under the more distant control of stored associations. The linearity of the temporal chain is unaltered; it is merely elongated to include previous associations. Nonetheless, this is an important historical advance in theorizing because it included the mind and implicated a whole other set of therapeutic techniques, later to be called cognitive therapy.

Jung and Adler, by contrast, emphasized the subjectivist side of Freud in their theorizing. As Freud's erstwhile associates, they had immense respect for the past and included it frequently in their respective individual therapies. Still, their use of the past was decidedly different from objectivists like Dollard and Miller and later behaviorists. Chapter 3, for example, shows how Jung relied heavily upon his construct of *synchronicity*. Although the superficial meaning of this construct is "meaningful coincidence," Jung made clear that synchronicity requires the subjectivity of time. Important "synchronistic" relations do not require temporal sequence or duration. Events can be together (contiguous) or apart—in space or in time—it does not matter. What matters is whether the person perceiving the events relates them together meaningfully.

Such theorizing has rather dramatic implications for individual therapy in Jung's view. First, the primary focus of therapy is the present, rather than the past. The "now" is where meaningful relations are formed, whether the elements of the relations are thought to occur in the past, present, or future. If anything, Jung stressed Freud's more telic side, seeing humans as primarily acting for the sake of their perceived futures (in the present): "Life is teleology *par excellence*; it is the intrinsic striving towards a goal, and the living organism is a system of directed aims which seek to fulfill themselves."[14] This was partly in response to Freud's objectivist version of libido. Freud saw the libido as an energy that flows continuously across time. Jung, on the other hand, referred to the libido as "horme," or the motivation that stems from valuing certain goals over other goals.[15]

Jung also avoided the determinism of linear accounts. He did not feel that neuroses began in the first five years of life. Therapists should not only focus patients on the present, therapists should also dissuade their patients from fleeing into the past. Patients unconsciously concoct fantasies and project them onto past events to avoid responsibility for the present. After one of his own patients "recalled" a childhood incident with her father, Jung remarked that "nothing is less probable than that the father really did this. It is only a fantasy, presumably constructed in the course of the analysis."[16] Jung's archetypes—those intrapsychic legacies of our ancestors—are treated similarly. They also are "constructed" for use in the present. They are not linear causes but instead "a priori categories of *possible* functioning."[17] To give special value to the past as the cause of the present is to deny persons their responsibility for their meanings and to move farther from, rather than closer to, the cause of the problem.

Other subjectivist and non-Newtonian characteristics are evident in Jung's approach to psychotherapy. Jung is so well known for his revolutionary theoretical insights, it may seem contradictory to view his approach to therapy as "atheoretical." Nevertheless, Jung did question the presumed universality of theory. He felt that it is vital to approach each patient uniquely and often cautioned against falling prey to theoretical rigidities: "Learn your theories as well as you can," Jung once said, "but put them aside when you touch the miracle of the living soul."[18] Jung practiced a *hermeneutic* form of treatment that calls for the abandonment of "all preconceptions and fixed ideas."[19] Thus, unlike Freud, Jung did not approach his patients as an expert. He gave them equal status and even claimed that the therapist is "in therapy" along with the patient.

Many of the same subjectivist themes can be seen in Adler's approach to individual therapy: here-and-now focus, rejection of linear determinism, recasting of the past, and atheoretical pragmatism. Two related elements of Adler's ideas are particularly striking in contrast to objectivist theorizing: teleological motion and contextual holism. As to the former, one of Adler's grounding assumptions was that all sentient beings are in "motion." Unlike plants that are unable to move, sentient beings follow the *law of movement*—always in process and following a line of action. At this point, it is tempting for an objectivist to presume that this law of movement is at least metaphysically similar to Newton's laws of motion—i.e., begun with antecedent forces, continuing through linear action and reaction across time, and so on (see Chapter 1). However, nothing could be further from the truth.

Adler's conception of motion is distinctly teleological and *not* mechanical.[20] That is, it is not based upon action and reaction across time, nor does it require an antecedent power to propel it. Rather, as Adler noted, "this power is *teleological*—it expresses itself in the striving after a goal, and in this striving every bodily and psychic movement is made to cooperate."[21] Unlike Newton's conception, motions are not pushed from the past in a mechanistic and physicalistic chain of events. They are instead pulled by a "future" goal as formulated by the mind: "This foreseeing the direction of movement is the central principle of the mind."[22] Moreover, Adler never allowed his students to think that teleology is underlaid by mechanism—that goals and direction are themselves determined by previous causes. Adler argued that telic direction is instead an originating cause, sufficient unto itself for explaining behavior.

Adler also resisted *reduction* of his conception of motion. Recall that Newton's conception led to the separation of parts of a process in time (see Chapter 1). Each portion of the whole could supposedly be studied independently as it occurred across the line of time. Adler, however, objected strenuously to this practice:

> It is . . . absurd to study bodily movements and mental conditions abstractly without relation to an individual whole. . . . The important thing is to understand the individual context—the goal of an individual's life which marks the line of direction for all of his acts and movements. This goal enables us to understand the hidden meaning behind the various separate acts—we see them as parts of a whole.[23]

With Adler, then, the parts that seem to exist independently across time actually do not exist independently. They participate in an underlying form or structure that gives them meaning, and they cannot be properly understood or studied without that underlying telic meaning. Another way to put this is that processes *across* time are secondary to (or derived from) processes *within* time (in the present). In contrast to Newton who took motion (and thus change across absolute time) as the primary reality of the world, Adler considered the "static" structures of the world as the basic reality and subjugated even his cherished "law of movement" to them.

Modern Individual Therapy and Time

Modern approaches to individual therapy continue to avoid Freud's mixed model. As the review below shows, *either* one side *or* the other of his mixed model is developed. Although a few practice subjectivist assumptions, the vast majority stress objectivist assumptions in their formulation of therapy technique. Even modern psychoanalysts—those who claim to follow Freud's original principles—tend to minimize his subjectivist constructs. It is as if our cultural objectivism and scientific Newtonianism are too strong for nonlinear and subjectivist ideas to survive. Subjectivist constructs are too difficult to grasp in a world which is "in" time; they are too "unscientific" in a world where everything must have antecedent causes. As we shall see, most subjectivist constructs are either dismissed or "linearized" into unidirectional and continuous forces. The few that have survived are herein examined for their anomalous properties.

Psychoanalysis

We begin by describing the modern linearization of Freud. Arlow represents this well in his statement about fundamental psychoanalytic principles:

> The psychoanalytic theory of personality is based on a number of fundamental principles. The first and foremost is *determinism*. Psychoanalytic theory assumes that mental events are not random, haphazard, accidental, unrelated phenomena. Thoughts, feelings, and impulses are *events in a chain of causally related phenomena*. They result from *antecedent* experiences in the life of the individual. Through appropriate methods of investigation [psychoanalysis], the connection between current mental experience and past events can be established. Many of these connections are unconscious.[24]

There is little or no room for temporal subjectivism in this statement. Even the "determinism" of Freudian teleology—i.e., being "psychically determined" by one's unconscious goals and wishes— seems ruled out by Arlow's "causal chain" and "antecedent experiences." Patients are viewed as having been programmed, in a very mechanistic sense, by past events, and are therefore caught in this causal chain forever. Indeed, according to Arlow, this "genetic approach is not a theory; it is an empirical finding confirmed in every psychoanalysis. In effect, it states that in many ways, we never get over our childhood."[25] If such statements are representative of this literature at all,[26] the linearity of time is not a debatable (and hence, a theoretical) issue. It is an established fact and a fascinating testament to the dominance of objectivism in psychology, especially since it occurs *in spite of* Freud's mixed model.

Freud's approach to therapy also seems problematic in this regard. If we cannot "get over" our childhood, as Arlow claims, then how do we help people in therapy? Arlow qualifies this by saying "in many ways," but how—given this deterministic revision of psychoanalysis and Freud's nondirective therapy approach—is there *any* way of getting over our childhood? Arlow goes on to describe several subtle, but substantive, modifications of Freud's therapy as the modern method of doing so. Here we see unacknowledged shifts from Freud's mixed model of psychotherapy to a more objectivist approach. Consider, for example, the crucial issue of transference in psychoanalysis. Freud viewed transference as a manifestation of the past as formulated from the perspective of the present, whereas for Arlow, transference is a "form of *memory* in which repetition in action replaces recollection of events."[27]

Nothing could be a clearer statement of an objectivist model of the mind. No subjectivist influences from the perceived present (moods, relationships) or future (intentions, wishes) are involved in Arlow's conception of transference. The mind is a repository of past events and transference is the reenactment of those stored events in the present. To "work through" such transference, reports Arlow, one needs to overcome "the amnesia for crucial childhood experiences."[28] From this objectivist perspective, then, insight is the remembering of previously forgotten childhood events. The meaning and reconstrual of these events—as part and parcel of the events themselves—is not mentioned. Unlike Freudian accounts, meaning is secondary to the supposed determinants of the meaning—events of the past.

Arlow's revision of Freud's assumptions represents many of the modern variations upon orthodox psychoanalysis. One exception may be the more relationally oriented variations, including some forms of ego psychology,[29] self psychology,[30] and object relations.[31] These probably come the closest to embracing a mixed model of time. Most relationally oriented psychoanalyses grow out of Freud's notion that "the relation to the external world has become the decisive factor for the ego."[32]

Certainly crucial to the modern development of this notion is the relation between the environment and the mind. If the environment precedes and is separated from the mind through linear sequence, then environmental factors are considered to cause and dominate the person (see Chapter 5 on Piaget). If, on the other hand, environmental and mental factors occur simultaneously, then neither necessarily holds causal sway over the other. A simultaneous relation could imply that both factors are influential in the present (and every moment thereafter). This would mean that simultaneous relations (in the present) are at least as crucial as sequential relations (from the past), and opens the door to a present-focused approach in individual psychotherapy.

Mahler, for example, explicitly rejects an objectivist blank slate approach to infancy for this reason (see Chapter 3). She disputes the usual external to internal (parental environment to infant mind) sequence of events and instead favors a reciprocal process in which infants, with their own unique capacities and endowments, play an influential role in all relationships. This form of theorizing is also reflected in the therapy practiced by those subscribing to object relations tenets. Just as the infant is seen to interact with significant others and not be passively determined by past events, so the adult in psychotherapy is seen to interact with others in ways that are not solely the product of past events, including past relationships. This allows the object relations therapist to deal with the here-and-now. Therapists can assume that present relationships, including the present therapist-client relationship, are as significant to mental disorder as past relationships.

Kohut strikes a similar theoretical pose. He is critical of most modern forms of psychoanalysis which base their theorizing on the objectivist assumption:

> that man's life from childhood to adulthood is a move forward from a position of helplessness, dependence, and shameful clinging to a position of power, independence, and proud autonomy . . . [taking] for granted that the undesirable features of adulthood, the flaws in the adult's psychic organization, must be conceptualized as manifestations of a psychological infantilism.[33]

Kohut recognizes that to conceive of infants in this weak and dependent manner is to place them at the whim of the environment (and ultimately the objective past). "Psychological infantilism" is the result—i.e., adults are viewed as the end-product of Newtonian forces operating upon them across time. Kohut opts instead for the assumption that the infant is "independent, assertive, strong . . . psychologically complete . . . no different from the adult."[34] This means that the infant and mother are both relatively equal elements of the interactional whole. Parents cannot and do not overpower the infant, as most objectivists assume. A true and reciprocally simultaneous interaction is possible.

The problem is that this type of theorizing has not captured the element of this interaction which is *not* a product of the environment. Theorists, such as Mahler and Kohut, are clear that there is such an element, but their explanations of it are vague. How can a mind have approximately equal power and independence to the environment *at birth*? How does one conceptualize a "psychologically complete" infant? Prominent Newtonian models of science do not facilitate answers to these questions, and a primarily objectivist psychology rarely attends to these subjectivist themes. Indeed, psychology texts typically linearize object relations theorists. Crider and his colleagues, for example, give a surprising amount of space to Kohut's theorizing in their introductory text, yet they lend this theorizing a distinctly objectivist tone. They explain that Kohut's notion of the self:

> grows out of the mother's treatment of the child. If the mother responds to the child's behaviors warmly, the child experiences him- or herself as joyful, happy, and worthy. If the mother is rejecting, the child experiences him- or herself as unworthy and empty.[35]

The quality of this description is clearly external to internal and past to present in linear flow. Subjectivist elements and the mixed-model nature of Kohut's theorizing are lost. Part of the reason is that Kohut and other mixed–model theorists are not sufficiently clear about their subjectivist elements. Without such clarity, their mixed model type of theorizing remains vulnerable to a linear interpretation by a primarily Newtonian audience. Moreover, any subjectivist components of their therapies (e.g., here-and-now orientation or current relationship focus) are similarly vulnerable to misinterpretation.

Behavior Therapy

Behavior therapy, as employed in its modern forms, encompasses many approaches to individual psychotherapy that would shock more traditional positivistic behaviorists. Rarely do behavior therapists stick strictly to procedures emanating from observable factors (e.g., conditioning). Techniques derived from operant and classical conditioning are still widely used, but most behavior therapists have become more liberal, employing the newest variations of cognitive and mediational strategies. The discussion begins (below) by documenting the obvious objectivist assumptions of conditioning approaches to therapy and then turns to popular cognitive therapies, which still retain the linearity of Dollard and Miller's historic mediational approach to the mind.

G. Terence Wilson, a leading behavior therapist, offers a helpful conceptualization in his 1989 review of the field. He delineates four main approaches to behavior therapy, each differing in the extent to which it uses cognitive concepts and procedures. At one end of this continuum is *applied behavior analysis*. This approach follows a strict Newtonian positivistic tradition in concentrating solely upon observable behavior and rejecting "all cognitive mediating processes."[36] At the other end of the continuum are the social learning strategies implemented by *cognitive and rational-emotive* therapists. Whereas applied behavior analysts rely upon Skinner's "radical" behaviorism, cognitive therapists tend to extend many of Bandura's theories to therapy (see Chapter 3).

Despite these theoretical differences, Wilson insists that all behavior therapy approaches "subscribe to a common core of basic concepts."[37] First, all subscribe to a learning model of human behavior. Even the more liberal models involving cognitive processes can be boiled down to behavioristic principles. As he puts it, "The rationale behind all these methods is that covert processes [cognitive processes] follow the laws of learning that govern overt behaviors."[38] This rationale is particularly important for cognitive strategies, as detailed below. Many cognitive therapists make claims that, at first blush, seem different from their behavioral and linear roots.[39] The second "basic concept" to which all behavior therapists subscribe, according to Wilson, is a commitment to *the* scientific method. The word "the" is italicized because his usage makes clear that he means a fundamentally Newtonian view of science, replete with objectivity, universalist laws, and reductionism. Wilson, for example, stresses the stability and predictability of behavioral principles across time, as well as the analysis of any clinical problem into "components and subparts."[40]

Applied behavior analysis, then, attempts to reduce problematic behavior to its components and subparts. Behavioral processes are presumed to occur across time and thus are regarded as separable in time. The assumption is that behavior is a function of its contiguous (in linear time) consequences. Accordingly, treatment procedures focus upon altering contiguous associations through techniques designed to induce new temporal contiguities, such as operant conditioning.

Another linear element in nearly all procedures is repeated exposure (see also Chapter 5). In treating phobias and fears, for instance, repeated exposure to the fear-eliciting object or situation is considered "the necessary (and usually sufficient) therapeutic component."[41] Neither the therapeutic relationship nor the patient's emotional state is important to success. These are unobservable processes and thus outside the domain of science and therapy.

Behavior therapy's ties to a Newtonian temporal framework run deep. For example, Wilson poses a "solution" to the common criticism from humanists that behavior therapy treats people as though they were not "free, self-directed agents."[42] As described in previous chapters, such freedom requires the ability to diverge from an objective past. If this past determines the person, he or she cannot make any novel contributions to the present stream of behavior, let alone exert any freedom of will. How then can a theorist who is tied to the "laws of learning" posit that a person can be free and self-directing? Wilson answers this by asserting that the environment and the person *interact* to produce the behavior observed. The environmental situation is not the sole cause of behavior; "person variables" also play large roles.

Wilson does not appear to recognize, however, that his answer merely begs the question. Person variables can themselves be products of the objective past, including past environmental factors. In fact, Wilson holds that "person variables are the products of each person's social experience and cognitive development that, in turn, determine how future experiences influence him or her."[43] In this case, the person/situation interaction to which Wilson refers is the sequential interaction between the immediate environment and the environment as stored in the mind. No real deviation from the linear past has occurred, nor is any deviation possible in the humanistic sense. The fact that Wilson views this as an answer to the humanist's criticism shows, once again, the grip of Newtonian temporal assumptions upon the discipline of psychology.

Cognitive Therapy

Wilson's comments are also important because many cognitive therapists claim to meet humanistic and phenomenological concerns. Indeed, many cognitive therapists contend that they are intellectual descendants of Immanual Kant and other phenomenological theorists and philosophers.[44] Although this may be true in certain respects, it does not appear to be the case concerning assumptions of time. Both Aaron Beck and Albert Ellis endorse mainstream information processing models as well as Bandura's social learning theory when formulating their cognitive techniques. As discussed in Chapters 3 and 5, most, if not all, of these models require Lockean, objectivist assumptions.

Furthermore, Beck and Ellis make explicit their reliance upon conventional learning theory, though they do so within a nature/nurture interaction. As Beck and Weishaar put it, "Within the constraints of one's neuroanatomy and biochemistry, personal learning experiences help determine how one develops and responds."[45] Beck holds that "people . . . respond to specific stressors because of their learning history."[46] In addition, cognitive structures as well as fundamental beliefs and assumptions "develop early in life from personal experiences."[47] Ellis is not substantially different. Ellis also regards the mind as mediational. His A-B-C theory views the A (activating event) as preceding the B (belief) and the C (consequent).[48] Although he takes pains to say that A does not directly cause C, the fact that it is a part of the linear chain is unmistakable. Moreover, the origination of the B, or beliefs of the mind, is a nature/nurture interaction similar to Beck's.

In discussing therapeutic technique, however, Beck and Ellis represent an interesting tendency in the applied disciplines of psychology to be somewhat inconsistent with their theory. That is, objectivist applications in therapeutic technique sometimes do not follow from objectivist theories. Although both Beck and Ellis formally endorse empiricist conceptions of therapy, based upon contiguity and repetition over time, they also advocate a style of therapy that only seems to fit a more subjectivist conception of the patient. Beck, for example, stresses a present-focused and collaborative style of therapy. He even discusses "voluntary" thoughts as though these can transcend deterministic factors of the past. Both Beck and Ellis tend to talk as if cognition has the power of "creating" and "reconstructing" the present. This would seem to argue for some type of escape from linear causes, yet such an escape is contradictory to their above assumptions. It may be that Beck and Ellis sense the need for this in their therapies but have no means of conceptualizing this in a primarily Newtonian psychology.

Existential Psychotherapy—Temporal Anomaly

This does not appear to be the case with existential psychotherapy. All the themes inherent in Freud's subjectivist side, especially as formulated by Jung and Adler, are employed in this approach to individual psychotherapy. Indeed, one of the originators and primary advocates of existential therapy, Medard Boss, was a member for many years of a seminar led by Jung. Existential therapists are not the only ones to capitalize upon subjectivist assumptions. The "person-centered" strategies of Carl Rogers as well as the gestalt techniques of Fritz Perls also manifest many subjectivist themes. However, both of these schools of thought and their founders acknowledge their existential roots. We therefore direct our examination of subjectivist assumptions in individual therapy to the root itself—existentialism.[49]

One of the founders of the existential movement in psychotherapy—Ludwig Binswanger—held that it "arose from dissatisfaction with the prevailing efforts to gain scientific understanding in psychiatry."[50] Not surprisingly, much of that prevailing "scientific understanding" construed the person as having been shaped or driven by forces occurring across time. That is, the philosophy of science at play was Newtonian. Existentialists, such as Rollo May, however, asked, "Where was the actual, *immediate* person to whom these [forces across time] were happening? How can we be sure that we are seeing patients as they really are, or are we simply seeing a projection of our own theories *about* them."[51]

May's statement is representative of the existential rejection of Newton's temporal framework. With perhaps one exception, existential therapists challenge virtually every temporal characteristic embraced by Newton and objectivist theorists in psychology. They do so not by advocating a new set of techniques or a new system of psychotherapy. Existential psychotherapists do not play an expert role in treatment, nor do they propose a universalized set of strategies or rules to follow. They do so by advocating a new philosophy of human nature. As May puts it: "Existential therapy is a paradigm, a frame of reference; it is not an organization with well-delineated rules. Hence, everyday arrangements of practice cannot be satisfactorily described."[52] Consequently, the existential frame of reference is stressed here.

Irvin Yalom, for example, in a comprehensive text entitled, *Existential Psychotherapy*, immediately differentiates an existential frame of reference from Freud's objectivist assumptions.[53] He shows how Freud relied upon time to decide which psychological issues were the more fundamental. As Yalom notes, Freud's conceptions of fundamental or "deep," "are to be grasped *chronologically*: each is synonymous with 'first.' "[54] In other words, the earlier that a conflict occurs in a client's life, the more significant the conflict is to the problem and treatment. According to Yalom, however, an existential therapist does not use linear time to distinguish deep or important issues:

> There is no compelling reason to assume that "fundamental" (that is, important, basic) and "first" (that is, chronologically first) are identical concepts. To explore deeply from an existential perspective does not mean that one explores the past; rather, it means that one brushes away everyday concerns and thinks deeply about one's existential situation. *It means to think outside of time.*[55]

For an existentialist, then, significant or deep issues are those that relate to the ultimate concerns facing the individual *at that moment.* "The *immediate, currently* existing ground beneath all other ground is important from the existential perspective."[56] Ultimate concerns or "grounds" include those related to death, meaninglessness, isolation, and freedom. As an example of how the existentialist thinks "outside of time," consider Yalom's view of the presentness of death. Unlike more linear thinkers, he does not view life and death as sequential and thus separable entities. In Yalom's words, "Life and death are interdependent; they exist *simultaneously*, not *consecutively*; death whirs continuously beneath the membrane of life and exerts a vast influence upon experience and conduct."[57]

Life and death are obviously more than physical conditions for Yalom. They are *meanings* that require a simultaneous existence for either to be fully comprehended. Life and death, in this sense, give meaning to each other. Knowing that we could die at any moment heightens our awareness and valuing of life. Indeed, the very meaning of life can be lost entirely unless its boundaries and limits are firmly grasped.[58] If problems in life occur, one of the potential "deep and significant" issues that a therapist should consider is the ultimate (and simultaneous) concern about one's mortality. To "think outside of time" in this way is to place emphasis upon *logical* rather than *chronological* significance. That is, existential therapists consider the important issues to be those that logically underlie, rather than come before, problems. Existential therapists do not attempt to excavate the past but instead direct attention to factors occurring simultaneously in the here and now.

One important set of these factors is the person and the environment. Unlike cognitivists and other types of temporal objectivists, persons and their environments are not separated for the existentialist, either in space or in time. The subject (the person) and the object (the environment) are considered to be a whole *within* time, i.e., a being-in-the-world (the hyphens denoting this irreducible quality). The environment in this framework does not occur "first" and "then" becomes cognized and stored by the person. Both person and environment exist and complement one another simultaneously. This is because persons are not "in" a linear and sequential world, existing independently of one another. Newton's framework is considered only one way of viewing change. As the biologist J. von Uexkull notes, "There is not one space and time only, but as many spaces and times as there are subjects."[59]

Each person constructs his or her own meaningful world, including metaphysical principles such as time and space. The same past or present circumstances can mean very different things to different people. The past is important, but it does not determine the present. According to May and Yalom:

> [The] world includes the past events that condition one's existence and all the vast variety of deterministic influences that operate upon one. But it is these *as one relates to them*, as one is aware of them, molds, and constantly reforms them. For to be aware of one's world means at the same time to be designing it, *constituting* one's world.[60]

Past or present meanings, in this sense, can never be divorced from the observer. The adjectives of many Eastern languages, notes Suzuki, include a "for-me-ness" implication. That is, descriptions of "objects" are never construed as objective and independent of the describer. To say that "the tree is tall" is to imply that "the tree is tall *for me*." This "for-me-ness" implies that the description—however objective it may seem—can be different for someone else. Indeed, there is no objective slant from which to view objects, or, for that matter, the "objects" of our memories. To say that "my mother was harsh" is to imply that "*for me* (in the present), my mother was harsh." No objective rendering of the memory is possible, even for physicalistic description. To return to one's elementary school, for example, is often to be astounded at its diminutive size relative to our remembrance.

Existentialists are also concerned about the future and its influence upon the past. May and Yalom evidence this in this passage: "Whether or not a patient can even recall the significant events of the past depends upon his or her decision with regard to the future."[61] Here Freud's teleology seems to play a large role. The formulation and reformulation of goals and intentions (in the present) are thought to govern *whether* one has memories and *how* one views memories. Memories that are irrelevant to one's goals, for instance, may not exist for all practical purposes.[62]

This reversal of the usual linear flow—future meanings governing past memories—is difficult to fathom from a Newtonian perspective. Many Newtonian theorists in psychology undoubtedly want to note that some object (or past) *has* to have been input; otherwise, there would be nothing with which to construct a future meaning. The problem is that this assumes a subject/object dichotomy, and objects (and the past) are not separate from subjects (and the present) for an existentialist. Subject and object (past and present) are a simultaneous whole, and the only world to which we have access is the "world of meaning." A memory, in this sense, is a different sort of present meaning, *not* a stored and separate object from the past.[63]

Existentialists do recognize the importance of lived time (or temporality) to life in general (see also Chapter 10). Although they have attempted to theorize "outside" of objective time, they have long valued and explicitly discussed lived and experienced time— time as meaning. As Bergson declares, "time is the heart of existence."[64] However, to separate our consciousness from time (as Newton did with absolute time), let alone to subjugate our existence to the constraints of some objective conception of time, is to lose our genuine existential relation to ourselves.

Truly meaningful events have little to do with "quantitative time." May and others have shown, for example, how love can never be measured in the number of years one has known the loved one.[65] Minkowski also proposes that the schizophrenic's "distorted" view of time is not so much an effect of his or her delusions as the delusions are an effect of the schizophrenic's view of time.[66] "Clock time" is obviously involved in our existence as people sell their time and regulate their day. Nonetheless, the issue here is the "inner meaning" of events. Humans may choose to conform to the shared meanings of clock time, but this does not imply that such temporal conventions govern these choices.

In fact, existential therapists highly value the transcendence of such conventions. Humans are understood to be transcending at every moment: "Existing involves a continual emerging, in the sense of emergent evolution, a transcending of one's past and present in order to reach the future."[67] May holds that there is a "neurobiological base" for this capacity. He points to Kurt Goldstein as describing a capacity for orienting oneself beyond the immediate limits of a given time and space.[68] Indeed, this transcendence of conventional time is vital to mental health. In Goldstein's studies, the loss of this capacity characterized the brain injured and mentally disordered patients.[69] They had difficulty transcending their past and present to see the possibilities of their future. "Their world space was shrunk, their time curtailed, and they suffered a consequent radical loss of freedom."[70]

Freedom is perhaps the most significant form of transcendence for the existential therapist. Freedom typically implies that a choice is the "starting" point of a pattern of behavior rather than the ending point of a cause-and-effect chain. To be a starting point, then, the choosing must somehow be conceptualized outside the flow of linear time. A Newtonian framework assumes that some event (or events) always precedes a choice. While such a sequence does not always imply causation, it does imply that these events must be part of a consistent and continuous "flow." Moreover, many psychological theorists routinely view all psychological processes as subject to linear causal chains of determinism.

If this were true, of course, a choice would not really be a choice at all; it would be the end-product of a causal chain. Without real choice, responsibility for one's self, one's actions, or one's meanings is illusory. Needless to say, many existentialists resist such conceptualizations. People may not be aware of all their choices for living and may have lost touch with their contribution to their existence. Nevertheless, people *do* have a free choice and *do* contribute to the meanings of their circumstances and their pasts, according to existentialists. Consequently, no circumstance or past can ever be totally responsible for these meanings.

Existential therapists also explicitly challenge temporal universality. As mentioned above, the expert role regarding patient difficulties and the formulation of treatment techniques are specifically avoided. The reason is, as Sartre put it, "Existence precedes essence."[71] That is, the essence of things—their truth—is dependent upon the existing person, "existing in a given situation (world) at a [specific] *time*."[72] The "truth" about a patient's problem is vitally related to the *context* of the problem. Existential therapists question therapeutic "essences" or universal principles that claim to be independent of a particular person's existential context. They are suspicious of diagnostic categories (e.g., depression) or universal recipes for treating all or even many people. Essences require abstractions that leave the concrete and unique individual out of the therapeutic picture. For the existentialist, there are as many different forms of depression as there are people who experience depression.

Actually, contextualism can be further extended. Contextual differences can be said to occur not only across persons but also across the life of a single person. In this sense, not only are there many forms of depres*sive,* there are also many forms of depres*sion* in the same person experiencing depression. It is not unusual for a so-called depressive to experience periods (however brief) of joy or peace. Unfortunately, such discontinuous flashes are often over-looked. As Newtonian "continuists," many therapists "see" *categories* of consistent behavior and emotion (attributions, traits) and overlook the moment-by-moment changes of the *patient* in front of them. Even the patients themselves are likely to construe their own experiences in categories of continuity across time, missing their own discontinuous "leaps" into other emotions and attitudes. This is unfortunate from the existential perspective because such changes can be mined for their therapeutic gold: *Why* was there momentary joy, and *how* did the patient move from his or her depressive stance in that moment?

Contextualism can really be extended *even* further—beyond most existential frames of reference. In a very significant sense, existential therapists still embrace a form of universalism and re-ductionism.[73] Their emphasis upon individual therapy is evidence of this. In fact, most existential psychotherapists have no problem treating individuals in a setting divorced from their natural con-text—the therapist's office. If existentialists are truly contextual, could treatment be so easily divorced from the context of a person's everyday life? Would not holism and contextualism argue for treat-ment of the *context* itself, at least occasionally? Yalom is one of the few existentialists to advocate group therapy (which provides a therapeutic context), but even he rarely advocates any sort of di-rect therapeutic contact with significant others in the patient's life. This is in stark contrast to many forms of family therapy (see Chapter 8).

Why this focus upon the individual? The answer lies in the subtle reductionism of most existential theories of personality. In-dividuals are viewed as having *within* them conflicting forces that require therapeutic remediation. Although this "inner struggle" is different from the inner struggle of the Freudians,[74] it is still an essentially internal and individualistic struggle, deemphasizing, if not obviating, the wider interpersonal context of the present. May illustrates this well: "When [existentialists] speak of the 'psychodynamics' of an individual, we refer to that individual's con-flicting, conscious and unconscious forces, motive, and fears."[75] Yalom also tends to see the individual as a self-contained structure for experiences. Death anxiety, for example, is "instrumental in shaping character structure."[76] That is, the individual is thought to house a character structure which is, in some way, shaped over time and "carried" across temporal contexts in a significantly universal sense. From this perspective, therapists need only involve the individual in facilitating therapeutic benefits.

Conclusion

Individual therapists—regardless of their theoretical stripe—reveal one or more aspects of a Newtonian temporal framework for explanation. Being satisfied with seeing an individual in psychotherapy reveals a reductionism of sorts. Freud is probably responsible, in part, for this notion of a self-contained individual: prepared with and carrying from place to place and time to time the necessary ingredients for assessment and treatment. Still, some anomalous therapies have challenged this universalist perspective. Existential therapists have seemed to go the farthest in this challenge, advocating a type of relational holism. Yalom's conceptions, which culminated in his advocacy of group therapy, probably demonstrate this best. However, even here, a subtle emphasis upon self-contained problems remains significant.

The rich and comprehensive theorizing of Freud also facilitated two streams of modern temporal assumptions in individual therapy: subjectivism and objectivism. Essentially, objectivism considers psychological processes to be "in" absolute time and thus regulated by its Newtonian properties. Subjectivism, by contrast, considers psychological processes to be the generator of such linear properties rather than the effect. Freud proposed a unique mixture of these sets of assumptions which has rarely been extended into contemporary individual therapy. Only relationally oriented psychoanalysts have attempted this, though even they seem to find the explicit formulation of subjectivist elements difficult to articulate. Other "mixtures" of a sort have occurred. Cognitive therapies, for example, endorse therapeutic practices that appear to stem from subjectivist assumptions, but their explanatory models seem to be primarily objectivist in nature.

The vast majority of therapeutic strategies operate under objectivist assumptions of human nature. The first step in most objectivist strategies is to assess the individual's unique past in some manner. This is thought to provide clues as to how the past has shaped and molded the individual's actions and feelings in the present. Such molding is typically considered to include several factors tied to absolute time: contiguity, repetition, reenactment, mediation, and cognitive storage and retrieval. Treatment, then, entails some sort of remolding or reshaping of the person's thoughts and/or behaviors. Mediations and storage systems are altered through a modification of the contiguous and repetitious relations of the person's environment. Sometimes the patient is encouraged to be directly involved in this modification (e.g., self-reinforcement or thought-monitoring), yet the fundamental rationale for patient improvement remains the same.

Only a small minority of therapists consistently employ subjectivist views of psychological processes. Existentially related therapists—including some humanists and gestaltists—come the closest to this, though their actual numbers are quite small. Treatment conceptualization is considered to be "outside" of linear time. The patient's own world is valued, including its own unique metaphysical principles of space and time. The patient's past is important but viewed as co-present with other meanings and not necessarily determinative of them. If anything, the present and future are more determinative of the past than the reverse. Other therapists may use existential strategies, such as present-focused therapy or reconstrual of the past. But when formal explanation is attempted, the subjectivist assumptions that underlie these strategies are rarely recognized or understood. Most often, and similar to modern revisions of Freudian concepts, these strategies and their subjectivist explanations are linearized to fit a Newtonian scientific model of the world.

Chapter 7

GROUP THERAPY

Group psychotherapy has only recently gained general acceptance as an effective mode of psychological treatment.[1] As described in the previous chapter, individual therapy has dominated the therapeutic scene. Mental health professionals have viewed abnormality as an individual problem, so they have assumed that it was best treated individually (as in the medical model). Increasingly, though, this individualistic focus is being questioned. Although group therapy is still considered inferior to individual therapy in many psychological circles, therapists have been impressed with the social facilitation and economic efficiency of groups. The presence of peers seems to encourage personal change, and more patients can potentially be helped in less time.

The presence of more people in therapy, however, does not necessarily mean a different conception of therapy. Much like individual psychotherapy, group therapy has strong advocates of Newtonian temporal assumptions.[2] Emphasis is often placed upon linear and reductive therapeutic techniques. A person's unique *past* is considered to be primary to the therapist's understanding and formulation of treatment, and the *individual* (rather than the group) is frequently the therapist's focal point. Indeed, many group therapies are more accurately characterized as individual therapies with the rest of the group as a participative audience.[3] Because these therapists are concerned primarily with the past, they reduce the group to individual interactions, engaging each group member, one at a time, in an attempt to discern and remediate each person's individual history.

Group psychotherapy also contains strong anti-Newtonian elements. A brief history of group therapy (below) illuminates many of the reasons for these elements, but the essential fact of its emphasis upon *groups* provides many theorists with a means of escaping the pervasive linear assumptions of psychology. Many group theorists, for example, contend that the group has a wholeness which transcends individual therapy techniques. Holistic qualities of the group allow each member to move away from their individual pasts. Their present relationship to one another becomes the crucial issue. These approaches emphasize the here and now and consider the group as an emergent gestalt with enormous potential for discontinuous change, as opposed to a collection of individuals who are continuous with past ways of acting and interacting.

The chapter begins by examining the historic roots of holism and individualism in group treatment. How has this "split personality" of group psychotherapy resulted? On the one hand, there is strong dependence upon the linear, continuous, and reductive assumptions of a Newtonian framework for explanation. On the other hand, there are strong elements of alternate views. What types of *practices* are now promulgated from these various views? This chapter sketches the dominant Newtonian group strategies and then outlines those strategies spawned by alternative assumptions. The latter seem to be accepted as more mainstream than the non-Newtonian alternatives of other psychological subdisciplines. The reasons for this are explored.

Linear Time and the Study of Groups

The study of groups has actually been an interdisciplinary enterprise, concerning primarily sociology and psychology. Although sociology and psychology probably "discovered" the import of groups simultaneously,[4] the formal study of group dynamics owes much to early sociologists who proposed that society is the result of a fundamental solidarity among people.[5] Durkheim, for instance, asserted that social life is based on small primary groups, marked by their interdependency and strong group identification (e.g., families).[6] These small groups were thought to form the basic building blocks of the larger society. He argued that we must understand such groups if we are to understand the processes that sustain society, culture, and the individual.

Historically, psychologists and sociologists have differed quite dramatically in their views of small groups. As Steiner put it, psychologists have adopted a more "individualistic" approach to groups, whereas sociologists have adopted a more "group-oriented" approach.[7] For the sociologist, "the individual is presumed to be an element in a larger system, a group, organization, or society. And what he does is presumed to reflect the state of the larger system and the events occurring within it."[8] Psychologists, by contrast, have presumed that the group reflects the state of the *individuals* within it. The group-as-a-whole was assumed to be a topic for economists or historians, rather than psychologists. As Lakin notes, "[psychological] theories of social interaction were in essence theories of individual actions."[9]

Forsyth contrasts the differing perspectives of sociology and psychology in the early twentieth century by revealing their respective reactions to the concept of "groupmind."[10] This concept originated in sociology. Durkheim suggested that assemblages of people sometimes acted with a single mind. He felt that such groups were linked by holistic forces which often subjugated the individuality of its members to a group will. Many sociologists followed Durkheim's lead by endowing emergent properties of the group with their own reality. The group was not just a collection of individual minds, or even a community of minds united by a common bond or purpose. Some groups had their *own* mind. This mind not only transcended its individual members and their separate pasts, it also reflected back on the individuals within the group, discontinuously changing them.

Most psychologists, at this time, rejected the reality of such concepts. Floyd Allport, for example, argued that the phenomenon of groupmind was unscientific. In his 1924 text on *Social Psychology*, he represented the differing perspective of psychology at this time: "Only through social psychology as a science of the individual can we avoid the superficialities of the crowdmind and collective mind theories."[11] Allport felt that groups could be fully understood by reducing them to their individual members: "the actions of all are nothing more than the sum of the actions of each taken separately."[12] He resolutely concluded that psychologists should never study groups because they are not scientifically valid phenomena. Allport thus exemplifies a tradition of individualism in psychology that continues in many forms to the present day. Even if "social factors" are studied, the primary focus is their influence upon an individual's actions.

What are the reasons for this individualistic orientation in psychology? Why was this maintained, at least historically, despite trends toward holism in sociology? Doubtless many factors went into this mode of thinking—economic, political, and cultural. On the other hand, such factors were likely to have affected sociology to a similar degree. What factors were unique to psychology? A major one was psychology's philosophy of science. Science meant *Newtonian* science to psychologists at this time. At a minimum, this implied that scientific investigation was positivistic. Because the wholeness of the group could not be positively observed, it had to be denied as a factor worthy of scientific consideration. Interpersonal relations were invisible links between people, and thus were not viewed with the same importance as the individuals who could be seen. Such a philosophy often led to observables being viewed as causal to unobservables, in this case the observable individuals being causal to the unobservable whole of the group.

But this is only part of the legacy of Newtonian science. Psychology's individualistic emphasis is related in part to the reductive metaphysic of Newton. Individualism is a distinctly modern concept, arising almost simultaneously with notions of absolute time (see Chapter 6). In the medieval period—before the rise of the concept of linear time—the individual was part of the corporate body of the community. Individualistic conceptions such as self and privacy had little meaning in this communal context. In fact, an "individual identity" in the modern psychological sense would have been unwanted. One's personal security and sense of belonging depended completely upon one's immersion in the broader community.[13] If a separate and unique self were apparent, social alienation, if not ostracism, would have been the likely result.

The concept of linear time contributed to an increasing focus upon the individual. Persons became more and more understood in relation to themselves in the past than each other in the present. The metaphysics of the world shifted from a synchronic view of *people* in corporate relation to one another (within-time), to a diachronic view of *individuals* in linear relation to themselves in the past and future (across-time). Before such linear views, time was taken less seriously and people were rarely connected with their childhoods. They were understood more in relation to their present situation (e.g., family, community, and vocation). With the arrival of absolute time, however, all events across time were linked into linear arrays. People became viewed as an effect of their separate and unique "time lines." Indeed, the person's self or identity has now become almost synonymous with the cognitive accumulation of unique past experiences.

This view is particularly accentuated in psychology because of its self-consciously "scientific" outlook. Allport's comments above evidence this emphasis. Science arose partly by placing a premium upon causal relations, which means Newton's *linear* causal relations. To understand some singular event, one looks to the motions and events preceding it in time. The immediate context of the event is typically given short shrift, because a major portion of that context—time—is considered to be invariant. Even a *group* of obviously related events is treated similarly. A Newtonian metaphysic considers a group of events to be a process which itself takes place across time. This has two reductive implications: (1) the group of events can be reduced to its individual components across time (since the wholeness of the group does not exist all at once), and (2) the group itself is subjugated to its own past, much like that of a single entity. This understanding of the group also implies universality. Whenever basic natural processes of a group are discovered, they are assumed to be lawful in the sense of universal to all times and places.

Psychology's adoption of these "scientific" explanations has affected its theory and research of groups. In general, interpersonal relations *across* time (e.g., early relations with one's parents) are valued over interpersonal relations *within* time (e.g., present relationships in groups). As revealed in the previous chapters, individuals are considered the end-products of causal chains stemming from their pasts. To understand these individuals, their unique pasts must be uncovered by delving into the repositories held in their brains. Individual identity or "self" has come to mean this repository. Concurrent relations, such as the influence of other people in a group, are at best secondary to the self in this sense. Moreover, interpersonal relations among group members are themselves thought to take place across time. Group members interact sequentially, so no relational connections at the same time are especially important.

The History of Group Therapy and Time

Such theorizing significantly affected early formulations of individual and group therapy. Freud, for example, wrote his most important works during the heyday of Newtonian physics. This was also the time in which debates about group phenomena, such as groupmind, were in full swing. Needless to say, Freud's emphasis was the individual. In keeping with the reductionist trend of psychology and medicine, he wrote little about concurrent social influences.[14] His focus was the past, at least as conceptualized by the person.[15] He did not rule out synchronous interpersonal forces, but he clearly did not give them much weight. Consequently, Freud did little to spawn group therapy. Later psychoanalysts were ultimately to be the leaders of the group movement, but their group therapy would always retain Freud's individualistic influence.[16]

The man often credited with first applying psychoanalysis to groups in the 1940s, Alexander Wolf, made explicit his emphasis upon the individual.[17] Although he found many advantages of the group—including its economy and social support—he made plain his focus upon individuals. Why did he adopt this individualistic focus in group therapy? Wolf's rationale is quite Newtonian: psychoanalysis is concerned with the individual character of each person's past. To reconstruct the patient's character properly, the therapist must delve deeply into each individual's personal history. Although an advantage of groups is their symbolic representation of the patient's family of origin, a therapist cannot reasonably expect to handle several different pasts *simultaneously* (i.e., as a group), not to mention the resulting transferences and countertransferences from those pasts.

A significant, but perhaps short-lived, deviation from this individualistic emphasis in group psychoanalysis is the work of Wilfred Bion. A British psychoanalyst in the tradition of Melanie Klein, Bion argued that the essence of group work is the unified system. To focus interventions and interpretations on the individual is to deny the opportunities afforded by groups. Bion, then, developed his own approach to groups, referred to as Tavistock study groups. These groups are classically psychoanalytic in every way except that the leaders confine themselves to the "behavior" of the group-as-a-whole and never individuals. Although this method was to influence human relations training and social work in Britain,[18] it has been highly criticized for its "mass group" emphasis.[19] Again, the reasoning of the critics seems to be Newtonian. Emergent qualities of the group are discounted (e.g., "group ego"), while the supposedly greater reality of individual qualities is substituted in their place. As a result, Bion's original methods are thought to be rarely practiced, even in the Tavistock clinic itself.[20,21]

Early psychoanalytic groups also helped establish another Newtonian precedent for group therapy—universality. Universality in groups is the notion that groups are underlaid with timeless structures. As Forsyth puts it, "Just as physicists, when studying an unknown element, analyze its basic atomic structure rather than its superficial features, so group dynamicists look beyond the unique features of group for evidence of these basic structures."[22]

Psychoanalysts, of course, stress basic structures of personality that are lawful and predictable across time (see Chapter 3). When they apply their individual theory to group work, they presume that groups also "behave according to predictable psychological laws."[23] Once group therapists know these universal laws, they can assume the more traditional role in psychoanalysis and medicine of *expert*. In other words, assumptions of universality allow group leaders to believe that they possess basic (and unchanging) truths about groups.

This assumption has also led to another large category of group therapies—structured learning groups. Such groups are planned interventions that focus upon a specific interpersonal problem or skill. They presume knowledge of a universal structuring of the group that has benefits for all or most of its members. Although these groups are sometimes difficult to differentiate from simple educational groups, they have a long history within the group therapy movement. For example, in 1905 Joseph Pratt employed this approach to help his patients suffering from tuberculosis. He gathered them into groups to instruct them on basic hygiene and health. Pratt was a supportive and inspirational teacher who saw his patients seeking solace from one another in their common condition. He later forsook this educational method in the favor of more group psychoanalytic techniques. However, structured learning and "behavioral" groups have continued to flourish.[24]

Not until Kurt Lewin's entrance into the group scene of the 1950s was this Newtonian framework substantially challenged. Indeed, Lewin is probably the father of most anomalies to the Newtonian paradigm in group therapy. Lewin strived to move away from historic causation and linear time (see Chapter 3). His gestalt heritage led him to champion present holistic relations that can emerge into new creations, transcendent of their linear pasts. Time does not automatically connect persons to themselves in the past and future. According to Lewin, time is a set of discrete "momentary sections" in which relations of the same moment "cause" the entities within them. Things and people are governed more by what surround them in the present than by what precede them in the past. That is, a "field" of persons and things constantly surround individuals. This field does not interact with individuals sequentially; it interacts with them simultaneously.

With such theorizing, it was probably inevitable that Lewin advocate group therapy. In fact, he stated quite baldly that "it is easier to change individuals formed into a group than to change any of them separately."[25] Groups are interpersonal "fields" in which people can be led to benefit one another through their simultaneous relations. The emphasis is thus on the present. The relationships of the group members take place in the present, and so the present is the pivot of all change. Present interpersonal "wholes" permit members to move beyond their individual pasts, and so the moment-by-moment "field" is the determining factor. This means that the agent of change is the group rather than the therapist. Unlike psychoanalysis, members learn through their experiences with one another rather than through the therapist's expert individual interpretation. According to Lewin, the therapist's role is more that of facilitator of group cohesiveness and healing interaction than that of expert guide through an individual's past.

Carl Rogers advocated a similar approach to group leadership, though he also implicitly challenged another legacy of psychoanalysis (and Newton)—universal group structure.[26] With his "nondirective" approach to therapy, the therapist need not make assumptions of underlying group structure. Rogers felt that the members themselves possess the capabilities to structure the group productively. The group as a whole can find its own direction without any "expert" help from the group leader. Rogers even questioned the concept of group leader and felt that the therapist should not enter the group with preconceptions about how the group *should* go or even rely upon therapeutic techniques for leading the group. Each group is different and changing. So-called basic structures of the group are not helpful in the minute-by-minute group facilitation. The Rogerian therapist is concerned more with a "way of being" than with an application of universal principles.

The Current Status of Group Therapy

The importance of these historic trends cannot be overestimated in understanding the current state of group therapy. Newtonian and non-Newtonian approaches exist side-by-side in the field, though there is little doubt that Newtonian approaches remain dominant. Temporal characteristics such as linearity, reductionism, and universalism are quite common. The past is considered primary, the individual is the ultimate focus, and universal group structures are still assumed to govern outcome. Theorists such as Lewin and Rogers have spawned important anomalies, but they have captured only a small portion of the field as a whole. Of the three major forms of group therapy—psychoanalytic, behavioral or educative, and existential or experiential[27]—only the last evidences anomalous themes.

We begin by reviewing the more dominant forms here[28]—the psychoanalytic and behavioral/educative—and then discuss possible anomalies to these in the next section.

Psychoanalytic Group Therapy

Individual psychoanalysis continues to be the guiding theoretical framework of most psychoanalytic groups. Classical analysis, particularly as recently construed, focuses upon the influence of the past on current personality functioning (see Chapter 6). Experiences during the first six years of life are seen as the root of one's intrapsychic and interpersonal conflicts in the present. Most of these roots are held unconsciously and must be brought into awareness to work them through successfully. Many techniques for discovering these temporal roots are used: dream analysis, free association, and transference, to name a few. Transference is perhaps the most important of these. It refers to the individual's unconscious shifting of feelings and attitudes from the past to the therapist in the present. Psychoanalytic technique is designed to foster this transference, examine the problematic past, and ultimately restructure the individual's personality.

The psychoanalytic group is a direct extension of this individualistic model. In fact, Wolf's historic application of individual techniques to group therapy continues to exert a profound influence upon contemporary psychoanalytic groups.[29] Transference, free association, and dream analysis are all viewed in individualistic terms, with the remainder of the group considered facilitators of these individual techniques. Even the main advantage of group therapy for most psychoanalysts is individualistic in nature.[30] The group is thought to provide a better sense of an individual's early family history. In contrast to individual therapy where a patient can only develop transference toward one person—the therapist— group therapy allows a patient several transference "objects." Indeed, a whole family of origin is symbolically possible in a group. Although such transference occurs in the present, the focus is upon how these relationships illuminate the individual patient's early history.[31]

Leaders of these groups typically impose a minimum of external structure to permit this "regression" to early childhood. However, leaders are also required to apply their expert knowledge of individual personality structure to interpret the regression. This expert knowledge is assumed to be universal to an important degree. The specifics of a particular person's history vary, of course, but the basic structure of the personality does not. Universal developmental structures are thought to lead to universal group structures. Interpretation of *inter*personal structures is viewed as a straightforward extension of the interpretation of *intra*personal structures.[32]

Lakin considers such individual conceptualizations to be a "paradox in contemporary psychoanalytic group that has never been examined."[33] That is, why do so many group therapies base themselves upon a model expressly intended to address individuals? Most groups, for example, adhere to what has been called the "principle of shifting attention."[34] Therapists shift their focus from one patient to the next during a single session. Other group members may be involved, but the interactions are therapist-centered and focused upon one group member at a time. Although this may be therapeutically beneficial, it does not capitalize upon the "groupness" of the group for treatment. Group characteristics may be used in *assessment* to understand an individual's original family dynamics. Nevertheless, such characteristics are rarely considered a tool for *treatment*.

To understand the paradox of not using the "group" in group therapy, we must remember the linear history of this field. Groupness requires an understanding of simultaneous relationships. Groupness cannot result from a member (or even a dyad) occurring one at a time because all members are not occurring *at the same time*. Group relations can perhaps be discerned by observing sequential relationships, one at a time, but this is only because the observer's memory makes the successive relationships simultaneous. For the group itself to have groupness, there must be simultaneous relations among all its members. The problem is that a linear metaphysic does not allow for the significance or understanding of these relations. Only sequential relations are considered significant. Therefore, the paradox stems from attempting to use a linear metaphysic to comprehend a nonlinear phenomenon—groupness.

Among contemporary group therapists, Alexander Wolf is perhaps the most explicit about such theorizing. Indeed, he considers the term "group therapy" a misnomer, because this implies the treatment of ailing groups, when the focus of analytic groups is ailing individuals.[35] Although many aspects of Wolf's therapy have been modified since his pioneering efforts in the 1940s, the past-oriented and universalist aspects of his techniques remain strong. Wolf specifically derides holistic and present-oriented treatments.[36] He contends that a preoccupation with holistic relations, here-and-now interactions, and group level phenomena only serve to distract from the important work to be done—revealing and reworking the individual's past. The group-as-a-whole is only important as it evokes familial transference and thus allows for a deeper searching of an individual's past.

Recent departures from classical psychoanalytic ideas have been more social in emphasis. For example, several ego psychologists have placed greater stress upon psychosocial development (see Chapter 6). Has this led to a greater valuing of interpersonal relations in group therapy? The short answer is yes, though the dominant approach remains linear and individual. Margaret Mahler's object-relations theory holds that past interpersonal "objects"—those significant persons who were involved in meeting the individual's needs—shape current interpersonal interactions.[37] Her study of the interactions between mother and child has convinced her that these are reflected in each person's intrapsychic structure later in life. Although this greater emphasis upon past *social* relations *has* translated into more emphasis upon present social relationships in psychoanalytic groups, the primary focus still remains upon the individual's past—though it is psychosocial in nature.[38]

Educative and Behavioral Group Therapy

Psychoanalytic groups are not the only form of group therapy to employ these linear assumptions. Although some educative groups advocate a dramatically different therapeutic strategy, they also focus upon the past, the individual, and universal principles for structuring the group. Behaviorists, of course, have long championed the Newtonian vision of science. This has been widely acknowledged concerning their philosophy of science.[39] Newton's avoidance of theorizing as well as his delimiting of knowledge to observables have long been important historic influences on this brand of psychology. What has not been widely known is the behaviorist's adoption of Newton's metaphysic and temporal framework for explanation. Even cognitive behaviorists and social learning theorists have continued to affirm this adoption, and so most group therapies that assume an educative stance reflect Newtonian temporal characteristics.

The term "educative" can encompass many forms of group therapy. In his review of major group therapies, Corey notes that rational-emotive therapy, reality therapy, and transactional analysis all "share the basic assumption of group therapy as an educational process."[40] Individuals are involved in a teaching/learning regimen in which they acquire new perspectives, behaviors, or emotions. Although the content of what is "taught" may vary greatly from therapy to therapy, the *process* of these approaches is essentially educational in nature. Indeed, the content of what is communicated can even be nonlinear in nature (e.g., reality therapy), but if the process of therapy is itself educative, then it is likely that the educational techniques used are linear (and individual) in nature.

This is because educational techniques often depend upon "learning principles." Even the vocabulary here betrays its Newtonian roots and characteristics. The term "principle" implies a *universality* of concept and explanation that denies the significance of context or culture. Furthermore, the notion of learning is difficult for most Westerners to fathom, except as it occurs and accumulates across *objective* time. This is certainly true of so-called conditioning principles, but it is also true of cognitive explanations (see Chapter 5). Behavioral and mental problems are a result of a person's *linear* past. It is through time that such problems are unlearned, and through time that more appropriate behaviors and attitudes are acquired. Such changes only occur in a *continuous* and orderly fashion as the person is shaped or reprogrammed; no discontinuous jumps or shifts are expected. If problems occur in this process, the replacement behaviors and cognitions can be *reduced* to their individual components across time for easier retention.

Behavioral group therapy clearly has all these Newtonian and "educational" characteristics. In fact, a leading behavioral group therapist, Sheldon Rose, acknowledges that it is probably more an educational than a therapeutic experience.[41] Certainly, all the temporal characteristics of learning theory are evident. Regarding *universalism*, behavioral therapists assume that all problematic behaviors, cognitions, and emotions—irrespective of context or complaint—have been learned and can be modified by new learning. Behavior therapists also follow a standard series of procedures in conducting the group.[42] Meichenbaum, for example, proffers a standard regimen for teaching people specific coping skills.[43] This is a natural theoretical extension of the assumption that the therapist possesses certain universal laws of human nature. In fact, a major criticism of the behavioral model is that such procedures are often too rigidly applied. As Corey observes, the therapist may focus on the techniques of the training strategy at the expense of understanding the "meaning behind the individual's behavior."[44]

This focus, however, is quite justifiable from a behavioral perspective. Meaning is deemphasized, if not disallowed, by the

Newtonian metaphysic underlying behavioral therapy. Similar to their individual therapy, behavioral group therapists assume an *objective* approach to problem formation, and "meaning" is not considered objective. Observable factors are crucial because these are the only factors that can be measured and quantified through supposedly objective temporal and spatial devices. As Chapter 4 demonstrates, assumptions of objectivity in science typically require assumptions of objective time. To measure change, or even to know that such change has occurred, some absolute means of measuring time is assumed to be required. Hence, assumptions of objective time often result in a deemphasis of subjective meaning.

Behavioral group therapists also make the *linear* assumption that the cause of the problem is in the past. Maladaptive social skills, for example, have been *continuously* shaped over time into their present form. Behavioral change is considered gradual, requiring many trials across time to have any effect. Prominent methods of intervention, such as "behavioral rehearsal" and "feedback," rely upon these objective characteristics. However, unlike psychoanalysis, behavioral techniques are not designed to trace causal chains back to their points of linear origin. Many psychoanalysts attempt to "regress" persons back to the "roots" of their problems, but behavioral group therapists focus upon the more immediate reinforcers, stimuli, or cognitions that they feel maintain the behaviors. Of course, all these behavioral factors (and the maintenance process itself) are considered to happen across time. Behaviorists simply trade the immediate for the more distant past.

The only Newtonian characteristic remaining is *reductionism*, and reductionism is apparent in many formal and informal aspects of behavioral groups. Similar to psychoanalytic groups, the individual is again the focus. The formulation of objective individualized goals is thought to be crucial to the therapeutic process. Group goals are sometimes recognized. Indeed, some social skill groups are based on commonality of purpose. However, each individual is viewed as possessing a unique learning history, and specific needs take precedent over any group level phenomena. In addition, the therapist handles most difficulties in reaching these goals with some sort of reduction of the behaviors to be learned. If, for example, someone is having trouble learning a complex social skill (e.g., assertiveness), then the skill is broken into its components (parts across time) for individual learning. This allows for selective feedback as well as selective reinforcement. The assumption is that the reduced components can eventually be chained together into the more complex pattern of social skill.

The more broadly educative group treatments place less emphasis upon strictly behavioral techniques. Still, as noted above, their primarily educational thrust makes them heavily reliant upon one learning theory or another, and most mainstream learning theories depend upon a Newtonian framework (see Chapter 5). The possible exception, in terms of the educative group approaches of Corey, is reality therapy.[45] Reality therapy seems to be based upon an essentially nonlinear theory. Its focus is more present- than past-oriented, because the individual is thought to have some means of transcending determinants of the past. This allows patients to be responsible for their present actions, and truly choose their goals for the future. Nonetheless, the *process* of this therapy is directive and individualistic, and little of the holistic qualities of the group are employed for treatment.

Experiential and Existential Anomalies

Let us now turn to the group therapy practices of psychology's so-called third force. Corey includes in this third-force existential group psychotherapy, person-centered (or humanistic) group psychotherapy, and gestalt group psychotherapy.[46] With the possible exception of the gestalt approach (as explained later), these group therapies have really attempted to *be* "group" therapies. Unlike psychoanalytic and behavioral groups, they have attempted to capitalize upon the groupness of the group—interpersonal interactions. Their assumption is that these interpersonal connections—in the *present* as well as the past—are basic to the human personality. As existential group therapist, Irvin Yalom puts it, "we are at all times obliged to consider man in the matrix of his interpersonal relationships."[47] This conviction has led third-force group therapists to several *non*linear clinical practices.

Central among these is a holistic emphasis upon group cohesion. Other forms of group therapy are concerned with individual or even dyadic relations (e.g., therapist and patient), but a holistic emphasis gives the highest priority to the togetherness of the entire group. Yalom calls this togetherness "cohesiveness," but unfortunately this word has many meanings in group work.[48] In third-force approaches, cohesiveness means more than simply a therapeutic alliance or member rapport; it means that the bond among group members is such that meaningful, and even intimate, relationships can occur. It also means that group interactions are a primary tool of treatment. Gestalt groups have traditionally been less concerned with member interaction.[49] But even here, recent versions have begun to stress the importance of group cohesion and meaningful interaction.[50]

Bringing a group of strangers—the usual constituents of a group—to this cohesive and intimate state is no small task. Third-force therapists have found that a here-and-now focus is vital.[51] The here-and-now is the notion that therapeutic attention is confined to the spatially proximal (here) and temporally present (now) as much as possible.[52] Of course, most therapists agree that the *process* of any therapy—regardless of the techniques being employed—takes place in the here and now. That is, the therapy qua therapy is not occurring in the future or in some other location, it is occurring here and now. At issue is the *content* of the therapy session. Many third-force therapists contend that the content should also be here-and-now focused. Discussion should center upon what is happening *now* in the relationships, feelings, and thoughts of the members, including the therapist.[53] Yalom, for example, argues that the more attention paid to the here-and-now, the greater the group's effectiveness and cohesion.[54] Because the bonding of the group occurs in the present, any facilitation of that bonding or cohesion process is in the here and now.

Here-and-now group approaches enjoy a surprising popularity. In virtually every other subdiscipline, the objective past is considered integral to the most dominant explanations and practices. Somehow, though, third-force group therapists have become a major factor in this subdiscipline. For instance, Yalom's description of the here-and-now approach is one of the best-selling texts in the field. Apparently, a large number of group therapists feel it is possible not only to understand a patient's problems with*out* knowledge of the (linear) past, but also to treat those problems effectively with attention *only* to the present.[55] Yalom seems to acknowledge the theoretical revolution that this entails. He notes that most therapists presume that the "real" or "deepest" causes of behavior stem from the past and argues that the here-and-now is more than a matter of technical emphasis. It is a matter of a wholly different *metaphysic*. In fact, he suggests—as Kurt Lewin before him[56]—that group therapists move to a "Galilean concept of causality" which emphasizes "field forces" of the present rather than causal forces from the past.[57]

As described above, however, the field of group therapy has not made this move. Despite Yalom's suggestion, no theorist or therapist has advanced a wholly different metaphysic for group therapy. What, then, has fueled the popularity of the here and now?

The answer seems clear: therapeutic power and effectiveness.[58] First, as mentioned above, attention to the present is a crucial element of group cohesion. Because group relations only take place in the present, the here-and-now must be used to illuminate and transform them for therapeutic benefit. Second, the present is important to any therapy that values *experiential* learning. To focus on present emotions and interpersonal relations is to draw attention to one's direct experiencing of these emotions and relations. Focus upon the past is at least one step removed from these experiences. Third-force theorists believe that patients should not just "talk" about their problems (as in psychoanalytic groups) or work on their problems through simulation (as in behavioral groups), they should experience the real thing with one or more of the real people of the group.

Pivotal here is another holistic assumption—the social microcosm. This is the notion that problems from the "there-and-then" (outside the group and in the past) eventually make their appearance in the present relations of the group members.[59] In this sense, the group constitutes a microcosm of society (and time), offering group members and the leader important *living* data as well as a direct opportunity to modify problematic interactions. The assumption of a social microcosm works in the reverse if the problem is satisfactorily changed. Any meaningful change in interpersonal skill or relationship in the group transfers itself outside the group and reconstitutes the patient's past.

How is this possible? As Yalom observes, the conception of a social microcosm has a holistic premise: "the part is reflective of the whole."[60] The head of a stick figure, for example, reflects more than its circular shape. Its "headness" implies something about the meaning of the other parts, i.e., they are also body parts. Moreover, any change in a part can lead to a change in the meaning of the whole. If the stick figure's head is redrawn as a horizontal line, the figure takes on new meaning—perhaps that of a television antenna (see Figure 7.1). Although the redrawing may appear to happen across some span of time, its effect on the meaning of the whole does not. The other parts of the whole change simultaneously with our realization of the change in the "head." No longer is the bottom portion legs, but part of the antenna. Moreover, this change in meaning occurs without any force across time reaching this particular part to effect the change.

Analogously, a group of people in therapy is also part of the greater whole of society and thus reflects that society. Like any whole, each of society's parts reflects its relation to the other parts.

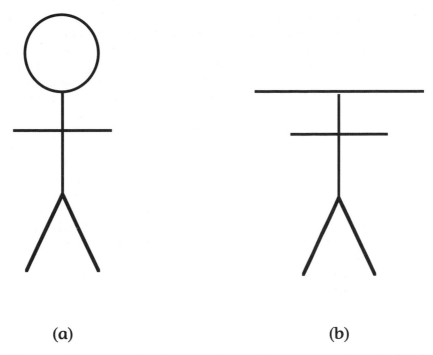

(a)   (b)

Figure 7.1   An illustration of simultaneous (or durationless) change in part-whole relationships

If group members change their holistic relations within the group, their relations to significant others *outside* the group also change. Such change is non-Newtonian because it does not occur across time—no elapse of time is necessary. Simultaneous with a meaningful change in relations within the group is a meaningful change in relations outside the group, though this may not, of course, be realized until later.

Although the metaphor of wholes is usually applied to issues of space (inside/outside), it is just as applicable to time. In the same way that the "here" of the group is reflective of the world outside the group, the "now" of the group is reflective of the past and future. Here, many third-force therapists seem to rely upon a Heideggerian conception of time. Heidegger conceived of the past, present, and future as all being wrapped together in the living now. Time is not a spatialized line, with an unchangeable past, empty present, and never experienced future; time is an ongoing dynamic whole. As our present changes, so do the meanings of the past and future, and as our pasts and futures change, so do the meanings of our present. In this sense, any meaningful change in the here and now of the group implies not only an instantaneous change outside the group but also a reconstitution of the person's past and future.

This type of explanation implies transcendence of linear determinism. Third-force group therapists are greatly concerned about transcendent abilities and their consequences.[61] Consider, for example, free will, which implies that persons can choose in opposition to their historical "determinants," though they can also choose in consonance with their past behaviors and influences.[62] This volitional ability is vital to a primary objective of these groups: helping members recognize their responsibility for the meanings of their lives. We cannot be a "victim" of our distant or immediate pasts. With the distant past, our childhood does not determine our present personality—we must assume responsibility for our present construction of this meaning. The past is the "servant, not master" of our psyches.[63] Likewise, no one in our more immediate past can determine our feelings. People in the group cannot "make" us mad, for instance, because our anger is determined by our interpretation of their actions, and we are responsible for this meaning in the present.

Transcendent capacities also suggest discontinuous change. If free will is possible (and not merely an illusion), then truly discontinuous changes can occur. Group members can change their usual pattern of interactions instantaneously. Although this typically does not happen without an awareness of this pattern and a cohesiveness of the group, the capacity to diverge from past interpersonal styles is assumed. Insight is another form of discontinuous change because it implies an abrupt change in one's attitudes or thought patterns (see Chapter 3). The group-as-a-whole can also evidence discontinuous change. As with any whole, an alteration in one part (or group member) can mean an emergence of a completely new identity for the group (and other members). Group therapists know all too well that one member's remark can send a relatively safe and cohesive group into a state of fear and fractionation.

This is one of the reasons that the expert role of a leader is deemphasized in these types of groups. These groups do not always follow a predictable course in therapy. Although commonality of the human condition can facilitate common themes of the group, such as fear of death and avoidance of responsibility,[64] the Newtonian "universality" of an underlying group dynamic is never assured. No two groups have to be alike, and no "laws" of the group are considered to govern therapeutic outcome. The group members themselves are thought to be the agent of the group structure, rather than any universal group structure being the agent of the group

members. Rogers, of course, is famous for this theoretical position—stressing the leader's role as "facilitator" of the group's
agency.[65] Mullan also makes clear that an existential method is a
*non*system regarding group dynamics.[66] Even gestaltists, with their
occasional imposition of group exercises, suggest these out of the
experience of their specific group rather than any presumption of
group universality.[67]

Conclusion

We have now reviewed the group therapy implications of the three
"forces" of psychology. At this point, we can see that the group
therapies of the first two forces—the psychoanalytic and behavioral—do not differ qualitatively from their individual therapy counterparts. Indeed, it is safe to say that these group therapies were
essentially developed to supplement their individual therapies. Psychoanalysts who employ groups typically do so to enhance their
individual therapy techniques. Chiefly, a greater number of transference objects may mean a better understanding of the patient's
early history. Behaviorists also extend their individual techniques
to group. The group is employed primarily to provide an easier
forum for individual learning. The presence of group members allows for the practice, feedback, and reinforcement of whatever training is being instituted.

In neither case, however, is the relationship among the group
members used as a treatment tool. Both, of course, are concerned
about rapport and the compatibility of members. This keeps any
negative relational factors (e.g., conflict) from disrupting the therapeutic regime. Still, neither attempts to facilitate or capitalize upon
the relationships that might develop among the group members
(e.g., intimacy). This may seem odd, given that the treatment is
intended to be a *group* therapy. Yet this reductionism is understandable considering the temporal assumptions employed. First,
group relations are not as "objective" as individual factors, at least
in the Newtonian sense of these terms. Second, the notion that
psychological events are arranged in a linear order gives more
weight to an understanding based upon individual histories. Third,
the universality of structures—whether they be personality or learning structures—makes relationships and "groupness" secondary, if
not irrelevant.

On the other hand, some group therapies originating from the
"third force" of psychology yield surprising anomalies to these tem-

poral assumptions. Not only is the relationship among group members *not* ignored, it is given top priority and holistic qualities. Evidence of this priority is seen in the bevy of non-Newtonian approaches that have arisen to facilitate healing interactions among group members. Chief among these is the here-and-now approach. Contrary to the supreme authority given the linear past in the explanations and practices of virtually every other field of psychology, these group therapists feel they can thoroughly understand and effectively treat patients by focusing almost exclusively on the present.

Some of the present qualities that bolster this assertion are the social microcosm, reconstrual of the past, and discontinuous change. The social microcosm permits all aspects of time and space to be centered in the here-and-now. As parts affect wholes (and other parts), so the microcosm of the present group affects past and future relationships outside the group. This also implies a transcendence of one's historical "determinants." Through human capacities such as choice and insight, new and original possibilities can be seen and acted upon, allowing for dramatic and sometimes even discontinuous changes. Of course, such possibilities limit the role of the leader. Neither the leader nor the underlying group structure can determine the course of the group. The leader facilitates healing interactions, but the group members remain the agents of therapeutic change and meaning.

Chapter 8

FAMILY THERAPY

Family therapy occupies a peculiar position in our survey of psychology's temporal assumptions. Most family therapists have long viewed "linear causality" as an improper and unproductive assumption.[1] The primary reason is its reductive quality. As revealed in the previous chapter, linear metaphors contribute to a focus upon the individual. People are viewed more in relation to themselves as individuals in the linear past, than in relation to each other in the synchronous present. This individualistic emphasis is considered contrary to the family therapy movement.[2] Family therapy gained its original impetus from the recognition that the present state of whole families was at least as important as the past state of individuals. In this sense, many would argue that the entire subfield of family therapy is an anomaly to the Newtonian temporal paradigm in psychology.[3]

This, however, grossly underestimates the power and subtlety of Newton's temporal framework for explanation. Actually, large portions of family therapy continue to embrace this framework.[4] Similar to group therapy, some family therapists still use individual personality theory to understand etiology and intervention. Although the context of the family is considered important, the individual is still the fulcrum of interventions. Other family therapists claim a "systems" orientation—seeing the family as a system of the whole. But even here, many therapists do not escape subtle forms of individualism and reductionism, as well as other characteristics of a Newtonian temporal framework.

As this chapter shows, the key to this reductionism is the mechanization of the system. Theorists view the family as analo-

gous to a cybernetic feedback loop where linear causality is replaced with "circular" causality. Circular causality considers the
succession of systemic events to circle (or "feed") back on itself.
Although this conception connects the various events of the system,
it still depicts them as happening one-by-one across time. Family
systems, in this sense, happen piece-by-piece across time. This
means that family theorists have retained the very assumption
that gave causality its linearity in the first place—linear time. As
long as the process takes place across time, the therapist or experi-
menter can only deal with it reductively as each piece reveals itself. Therapists who wish to intervene in the family can only do so
one member (or piece of the system) at a time. This form of family
therapy, then, is really successive individual therapy with each
member as the system occurs.

The field of family therapy also has its genuine anomalies. As
the history of this modality reveals, the root of systemic conceptions is twofold—mechanistic and organismic. While the mechanistic is clearly the more dominant at this time, the organismic has its
own modern manifestations. From the organismic perspective, the
family is a dynamic structure (or "organism") of simultaneously
interacting parts, without loops or circles across time. Although the
structure itself changes, this change depends upon one's point of
view and is not measurable through some objective standard of
time. Techniques of therapy are thus more spontaneous than planned
and focused more upon the present state of the system than upon
some deterministic past or idealized future.

This chapter begins by showing the historical roots of these
conceptions. It chronicles the struggles of family therapists to escape the effects of linear causal notions of psychotherapy and psychopathology. Although systemic and cybernetic conceptions are often considered synonymous, this review demonstrates important
differences concerning assumptions of time. Cybernetic conceptions
are then shown to dominate current family therapy theory, along
with the absolute temporal assumptions and characteristics of any
such mechanistic notion. The one exception may be second-order
cybernetics. Here, and in the atheoretical work of several family
therapists, systemic anomalies to the cybernetic and reductive formulations of family theorists are described.

The History of Family Therapy and Time

The movement of family therapy toward a *systems* therapy
probably began in the 1940s. Gregory Bateson, for example, dis-

cusses the confluence of ideas occurring in the mid 1940s that led to "what sort of a thing is an organized system."[5] Moreover, important conferences on circular causality and cybernetics were organized during this period, (e.g., the Macy Conference of 1946).[6] Of course, individualistic approaches to family therapy originated much earlier (as reviewed in Chapter 6 and 7). Psychoanalytic and behavioral approaches to family therapy are often thought to be more individualistic in orientation.[7] Systemic approaches, however, consider the family to be more than merely a context for individual behaviors and personalities. Systemic perspectives attempt to understand and treat the *entire* family as a whole.

As mentioned in the introduction to this chapter, there are essentially two types of systemic perspective: mechanistic and organismic. The mechanistic perspective sees the relational and interactive qualities of the system as happening across time. A definite linear sequence of systemic events is delineated, dependent upon the particular structural or "hardware" conditions.[8] Consequently, each component of the system is primarily reactive to the preceding component. An organismic perspective, on the other hand, views the interactive and relational aspects of the system without linear time and sequence. Interactions among parts are simultaneous rather than sequential. The system-as-a-whole is sometimes envisioned as occurring across time—growing and differentiating—but the wholeness of its parts (their relationships to one another) does not. Indeed, from an organismic perspective a whole cannot *be* a whole without the simultaneity of its parts.

The foremost representative of the mechanistic perspective in the history of family therapy is cybernetics. Cybernetics was formulated in the early 1940s to understand and account for self-regulation. Self-regulating systems are those systems that seem to direct and control themselves, rather than *being* directed and controlled by events external to them. Many machines (e.g., thermostats and engine governors), biological organs (e.g., the brain and kidneys) and human organizations (e.g., businesses and families) appear to evidence these characteristics. These systems do not seem amenable to the more linear notions of explanation, such as stimulus-response and linear causality. These systems appear to show some capacity for self-governance, and some degree of autonomy from simple cause-and-effect chains across time.

How then are such systems to be understood? Historically, mechanistic metaphors have always been close at hand. Norbert Wiener, perhaps the father of cybernetics, described his mechanistic understanding of self-regulation in this passage:

The word cybernetics is taken from the Greek *kybernetes,* meaning steersman. From the same Greek word, through the Latin corruption gubernator, came the term governor, which has been used for a long time to designate a certain type of control mechanism and was the title of a brilliant study written by the Scottish physicist James Clark Maxwell eighty years ago. The basic concept which both Maxwell and the investigators of cybernetics mean to describe by the choice of this term is that of feedback mechanism, which is especially well represented by the steering engine of a ship.[9]

The key for Wiener, as for most cyberneticists, is the notion of *feedback.* Self-regulating systems govern themselves as a result of feedback mechanisms. These mechanisms permit the system to "know" what it is doing and correct itself appropriately. As Wiener puts it, feedback is "a method of controlling a system by reinserting into it the results of its past performance."[10] By focusing upon engine governors and the like, he made self-governing *machines* the prototype of such feedback. No longer was the cybernetic machine one of several types of self-regulating phenomena; the cybernetic machine was *the* model of how all such self-regulating systems operated.

This also meant that linear time was endemic to all systems. Any systemic process involved in any system was to be viewed as happening along the line of time.[11] The process of feedback itself, as Wiener's definition shows, gives primacy to the past. Feedback, after all, is the reinsertion of the system's *past* performance to regulate its operation. Feedback cannot occur simultaneously with performance; it must follow or "circle" back on itself to be fed back. Indeed, this became an important epistemological distinction: systems operate according to *circular,* rather than *linear,* causality.[12] Instead of isolated and unidirectional chains of causes and effects, such chains "bend" onto one another in a circle. This notion was quickly recognized to have several advantages over linear causality. Nevertheless, the fact that such circular events occur across time was never questioned.

Through the early 1950s, cybernetic theorizing was primarily limited to engineering problems and computer applications. When circular principles were united with information theory, however, the use of cybernetics was greatly expanded. The flow of information in a human system was viewed as directly analogous to the flow of energy in an engineering system. Just as signals from a thermostat are continually passed through the various components of a temperature-control system, so messages in a family are continually communicated through its various members. Circular causality thus became the reigning vehicle of explanation. In fact, this "reign" was so influential that cybernetics became the preferred explanation for all types of human systems.

Gregory Bateson is often credited with a vital role in this move to cybernetic explanations. His translation of the ideas of engineering and mathematics into the language of behavioral scientists was crucial to the development of many family therapies. Bateson's studies—particularly his landmark study of the communications of schizophrenic families[13]—led him to question the adequacy of linear theorizing, especially conventional learning theory. When cybernetics came along in the work of Wiener and Von Neumann,[14] Bateson felt that he had found the alternative framework for families and cultures he required. His influence has been so great in the family therapy literature that general systems theory and cybernetics are now viewed as essentially the same.[15]

Nothing, however, could be further from the truth—at least according to the father of general systems theory, Ludwig von Bertalanffy. In his view, the cybernetic or feedback model "falls short of being a general theory of systems."[16] Bertalanffy acknowledged the importance of cybernetics as a *type* of system, but he clearly championed the *organismic* conception of systems as the more fundamental and the more applicable to human organization. Surprisingly, though, these points are almost never mentioned in family therapy texts. Bertalanffy's general systems theory is often noted, but its organismic and *non*feedback foundations are overlooked. Indeed, general systems theory is often *equated* with cybernetics, at least on fundamental issues. A leading family therapy text, for example, declares that "both [general systems theory and cybernetics] are built upon the same assumptions."[17] Dorothy and Raphael Becvar—in another text—draw similar conclusions in this passage: "Although considered less mechanistic than cybernetics, general systems theory is equally concerned with feedback mechanisms . . . and, in fact, there is little that separates the two theories from each other."[18]

This type of misconception seems to be a clear example of the dominance of the Newtonian framework in psychology. Family therapists have explicitly desired a nonlinear perspective from which to view the family system, yet the mechanistic framework of cybernetics has not only prevailed in popularity, it has succeeded in impeding the recognition of nonmechanistic alternatives. Bertalanffy himself seems to have found this to be a troublesome issue:

> The interest in these [mechanistic] developments is well understandable and deserved in view of the role cybernetic systems, computers, and "servos" of many kinds are playing in industry and modern life. Not infrequently, this has led to equating "cybernetics" with "systems theory." This, however, is a misunderstanding that needs correction.[19]

Bertalanffy went on to distinguish between mechanistic and organismic approaches in many of his books. Unfortunately, the explanatory power of Newtonian mechanism has subsequently blurred these distinctions in family theory. Therefore, let us attempt to clarify some of them here.

According to Bertalanffy, organismic approaches differ from mechanistic approaches because the former recognize the essential wholeness of systems. Mechanistic approaches are rooted in the "Newtonian simplification of one-way causality."[20] That is, cybernetic and feedback approaches to systems still belong to a Newtonian causal framework. Bertalanffy was not impressed with circular causation as a move away from Newtonian one-way causality. As he put it, "feedback regulation is by way of linear and unidirectional (although circular) causality."[21] In other words, feedback loops and circular causation are still linear and unidirectional in his view.

What can this mean? Feedback is obviously not unidirectional in terms of *space*. It moves back and forth or circles around through the various components of a system. Bertalanffy was quite aware of this and routinely described circular causal conceptions in this manner. His notion that circular causal conceptions are still "unidirectional" (or one-way) must then refer to *time*. That is, despite its circling in space, feedback still proceeds forward in linear time, one component being separated from the next *in* time. This meaning is especially clear when he discussed the differences between organismic and mechanistic approaches to interaction. Cybernetic theorists view each component of the system as happening in a linear sequence, so components "interact" in pairs—cause to effect—in what Bertalanffy called "two-variable problems." Organismic approaches, by contrast, consider all variables of the system *at the same time* in a constant state of "multivariable interaction."[22]

Although Bertalanffy went on to describe several other differences between these two approaches, it seems clear that a major one—if not *the* major one—involves the issue of time. Whereas cybernetic components invariably take place *across* time, organismic "components" take place at the *same* time. That is, cybernetic conceptions assume that components of a system occur in a linear sequence dictated primarily by its past. "Information" relayed by the previous component causes the present component—"reactive" rather than "active."[23] Organismic components, on the other hand, are always actively contributive. They are constantly "on" and constantly relating and relevant to the system's state at any point. Bertalanffy called this "dynamic interaction,"[24] and realized that such interactions make it possible for discontinuous changes. Unlike mechanistic regularity, some emergent qualities of the whole are unpredictable from its previous qualities.[25]

Bertalanffy also contrasted organismic and mechanistic approaches on their reductive qualities. For example, parts of a mechanistic process never appear (or work) as a whole at any one time; they occur piece by piece across time. This meant to Bertalanffy that the whole of the system (more than a two-variable problem) can never be dealt with or understood. This is one of the reasons he believed that cybernetics falls short of being a truly systemic perspective. It offers no insights into the *simultaneous* interplay of *all* the system's components. Organismic conceptions, by contrast, portray the system as a whole "within" time, i.e., at any one moment. From this perspective, no component of the whole can be understood without knowing its relationship to all the other components. No single component comes before it, and the present relations among its parts determine its properties more than any past state of the system. The system can be seen as having a past. However, no necessary relation to that past is required, because a new gestalt of the system's components can transcend the past at any time.

The Current Status of Family Therapy

The current state of family therapy incorporates both of these approaches to family systems, as well as the more traditional individualistic approaches of psychoanalysis and behaviorism. We will review the modern manifestations of each approach in turn, beginning with the more individualistic. As might be expected, modern manifestations of the organismic approach are the most clearly anomalous to the Newtonian paradigm in psychology. Still, individualistic and mechanistic approaches—though predominantly linear in nature—sometimes employ techniques that seem to violate Newtonian temporal assumptions. Part of the reason for this is a disparity between theory and practice. Even though a mechanistic theory (or "cybernetic epistemology") is explicitly endorsed, it may or may not be applied consistently in practice, especially as it concerns matters of time.

Variance between theory and practice is probably an issue for all forms of psychotherapy (see Chapter 6). Nonetheless, it may be a special issue for family therapy because mechanistic and organismic assumptions have been so confounded in this literature. For many theorists, cybernetics *means* systemic wholeness,[26] and thus the cybernetic tie to machine metaphors (and their separation of components across time) is vague at best. Certainly, this literature has blurred Bertalanffy's distinctions between mechanism and organism. As we shall see, this blurring has resulted in a mix of systemic techniques across the field, if not a mix of systemic assumptions within a single school of thought. After describing this mix in mainstream family therapies, a relatively new conception—the cybernetics of cybernetics—is outlined. This conception may hold some promise for escaping its mechanistic origins, yet its temporal roots are still Newtonian in nature.

Individual Models of Family Therapy

Many psychotherapists seem to consider any treatment conducted in a familial context to be family therapy. This leaves the field wide open for individual approaches. Therapists can view themselves as conducting family therapy while essentially applying their individual therapy training to each of the family members in turn. Some traditional psychoanalysts and behaviorists would fall into this category. Although they value the information provided by all or part of the family, their primary focus is the individual. They may attend to present "family dynamics," but the ultimate center of therapeutic attention is the individual in relation to his or her past, particularly transgenerational family issues.

A case in point is the theory and technique of Nathan Ackerman. Family therapy has long been divided between those favoring an intrapsychic approach and those favoring a systemic approach. "Ackerman was the most outstanding proponent of the [intrapsychic] position."[27] Problems stem more from conflicts and past experiences contained *in* the psyche than from interpersonal conflicts and present experiences stemming from *outside* the psyche in the family system. Ackerman did incorporate many systemic ideas such as family homeostasis. Nevertheless, as Michael Nichols observes, "he emphasized the intrapsychic effects of families on individuals more than the behavioral sequences, communication, and interaction that [mechanistic] systems-oriented therapists stressed."[28] With Ackerman's death, the family movement seemed to lean more toward a systemic orientation, though many from this first generation of intrapsychically trained therapists continue to favor a more individualistic orientation.

Murray Bowen can also be considered among this number. Despite some obvious systemic theorizing, he does not quite escape his own psychoanalytic training, particularly in his approach to treatment.[29] Bowenian therapists must think in terms of family systems when attempting to understand their dynamics, but such wholes are employed to gain greater insight into the individuals which form the family system. The focus, then, is individual intrapsychic functioning. The individual is the intrapsychic carrier of past family dynamics, which is then transmitted across time to the present complex of relations. On the other hand, Bowen is also concerned with present family relationships and theorizes well beyond the simple therapist-client dyad. What is the reason for this seeming paradox?

As Nichols notes, the answer lies in Bowen's emphasis upon the mental preservation of past relationships: "The emphasis is on those mental structures that preserve early interpersonal experiences in the form of self- and object-images."[30] In other words, present relations among family members are a concern because they serve as an important guide to the past relationships that formed them. Bowen shows how many complex family dynamics (e.g., triangulation) can be interpreted and even changed systemically. Nonetheless, "the Bowenian approach seeks to change families by changing individuals."[31] Thus, declares a prominent family therapy text "linear causality characterizes the etiology and treatment of problems defined in Bowenian family therapy."[32]

Behavioral family therapists manifest a similar emphasis upon the individual. Although Bowen and Ackerman refer to systemic concepts in their understanding of the individual, most behavioral approaches make no pretense of a systemic orientation at any level of theorizing. Indeed, theory itself is viewed as less than relevant, unless it is, of course, learning theory. Behavioral approaches to

family therapy rely—as their group and individual therapy counterparts—on a scientific base defined in the positivistic tradition of
Newton. This rather naturally places their family approaches in
the Newtonian temporal tradition as well. Sometimes behavioral
approaches are viewed as ahistorical, because behavioral assessment is concerned only with "current" determinants of behavior.[33]
Nevertheless, these determinants clearly occur across time. The
term "ahistorical" is intended more as a contrast with psychoanalysis which concerns itself with the more distant past.

The work of Gerald Patterson is perhaps best known among
the many behavioral approaches to the family. He and his associates have emphasized parenting skills as a vital component of family therapy. The usual goal of this therapy is to change the parent's
response to the child in order to change the child's behavior.[34] Parents are taught the skills, but the child is often construed as the
"identified patient." To know more about a child's problem, careful
behavioral assessment is necessary. Assessment is generally geared
toward finding lawful regularities that are presumed to exist in the
dyadic relationship between parent and child. These lawful patterns are then broken into their various parts. The parents must
discriminate between desirable and undesirable behaviors and ar-
range the appropriate consequence when the behavior is observed.
The therapist's job is to help the parents gradually shape the more
desirable behaviors into a more adaptive chain of behaviors across
time.

Parent-skills training thus exemplifies many Newtonian assumptions. First, the primary focus is upon the individual. The
problem is not in the family qua family, but in the individual's
behaviors as learned through time. As Becvar and Becvar note,
"Reciprocity in the parent-child relationship is recognized, but the
treatment focus is basically linear and the pathology or problem is
located in the child."[35] Second, the universality of behavioral principles is unquestioned. These lawful regularities exist across people
and across temporal contexts. And third, when these universals are
found, they are reduced to their separate components across time.
Although these components are later reconnected through a "shaping" procedure, the ultimate product is viewed as a linear chain of
behaviors across time.

Recent formulations of behavioral marital therapy have seemed
less linear and reductive. Consider the marital approach offered by
Jacobson and Margolin:

Since each spouse is providing consequences for the other on a continuous basis, and since each partner exerts an important controlling influence on the other's behavior, the marital relationship is best thought of as a process of *circular and reciprocal sequences* of behavior and consequences.[36]

This conceptualization does not appear to be as individually oriented as an approach that teaches a parent how to cope with a problematic child. Indeed, the emphasis here on "circular and reciprocal sequences" seems to imply relationships rather than individuals. Still, the fact that relationships occur through *sequences* means that, once again, the conception is that of one member at a time, instead of all family members interacting at the same time. If this conception is accurate, the entire marital system can never be treated at any one point in time. Moreover, each individual is an effect of the previous behavior of their spouse and thus a reactor to that behavior. Although it is true that the individual is also a cause of the next moment's behavior, this "causal" factor is little more than the relay of something before it in time. The individual who is the relay may add his or her own unique stamp to the relay (e.g., past experiences). Nevertheless, no family member can be an active factor in the sense of a new and original (not from the past) contribution to the system.[37]

Mechanistic Systems Therapy

Such circular and reciprocal processes bring us directly into modern manifestations of mechanistic theory in family therapy. Technically, all family therapies that claim a cybernetic or circular causal foundation would be considered "mechanistic" in conceptualization, at least according to Bertalanffy.[38] As mentioned above, however, the disparity between theory and practice in family therapy may lessen the mechanistic impact upon actual technique. Concepts of cybernetic sequencing and mechanistic homeostasis, for example, may not be followed consistently in treatment. Nonetheless, our focus here is the nature of the *explanations* given such treatments. In these explanations, many family therapies continue to rely upon a Newtonian framework.

Communication approaches to family therapy probably depend the most heavily upon cybernetic assumptions for their explanations and techniques. Recall in our historical review that cybernetic conceptions were initially brought forward from engineering. Their first application to human organizations occurred when cybernetic principles were wedded to communication theory. Instead of electrical or chemical energy being circulated through the feedback loop of a machine, information (communication) is circulated through the feedback loop of a family. Sis misbehaves so dad gets angry and then mom cries, and this leads to sis misbehaving again, and so on. Family therapies emphasizing this type of looping communication have thus held the closest ties to cybernetic conceptions.

Three major approaches to family therapy stress these ties— the Mental Research Institute (MRI), strategic, and Milan approaches. All explicitly claim cybernetic assumptions. Watzlawick, Weakland, and Fisch—leading advocates of the MRI model—assert that "the formulation that is perhaps most relevant to our subject matter is the one given by Ashby for the cybernetic properties of a machine with input."[39] Perhaps the most outstanding proponent of the strategic approach, Jay Haley, says that his "communication ideas are largely derived from Gregory Bateson,"[40] the historic promulgator of cybernetic principles in family therapy. In fact, Haley's mechanism is easily recognized in his notion that communication occurs in "bits," and that there are two (cybernetic) styles of communicating such bits: digital and analogic.[41] Mara Selvini-Palazoli, an originator of the Milan approach, considers it to be "derived from the models offered by cybernetics and communication theory."[42] This model is so important that she envisions individual elements of the system to be analogous to systemic "circuits."[43]

These cybernetic foundations are all considered to be "circular causal" in nature. The prototype of this circular nature is the following mother-daughter interaction: The mother nags and so the daughter defies her, but then the mother only nags because the daughter defies her. Such chicken-and-egg issues are common in families. From a cybernetic framework, the attempt to determine a linear cause and effect in these situations is futile, if not detrimental. The mother is not the cause of the daughter's behavior, nor is the daughter the cause of the mother's behavior. Any attempt to determine a linear cause results in blaming one or another member of the family. According to communication therapists, *both* family members are thought to be caught in a "reverberating system, a chain reaction that feeds back on itself."[44]

At first glance, such theorizing seems a far cry from a Newtonian framework. It is undoubtedly an advance upon simple linear formulations, but this circular "chain reaction" does not function as the cyberneticists might wish. It does not avoid blaming the individuals within the system, nor does it conceptualize the family as a whole. The individual's "communication" *is* responsible for the communication that follows it. Its circular nature does not prevent the antecedent event from being causal to the consequent event. The mother and daughter *can* blame each other for their behavior because each behavior *is* a reaction to the part of the chain that precedes it in time. Selvini-Palazoli agrees with Haley that the behavior of each family member "is controlled and influenced by the other members of the family."[45] Although it is true that such theorizing precludes the *distant* past from being a primary factor— i.e., the "chain" must eventually circle back on itself—this theorizing does not escape the linear time implication of blaming a reduced piece of the system for the system's state at any one moment.

This reductionism is also evidenced in formal definitions of the system. Systems of all types—whether dysfunctional or functional—are typically defined as "patterns of behavior that are repeated in regular sequences which involve three or more persons."[46] Like all Newtonian frameworks, any systemic process takes place across time. Paul Watzlawick captures this quality in this passage: "Implicit to a system is a span of time. By its very nature a system consists of an interaction, and this means that a sequential process of action and reaction has to take place before we are able to describe any state of the system or any change of state."[47]

When a system is spread across time in this manner, the task of the therapist is to "break into the repetitive but self-perpetuating cycle."[48] The problem is that any intervention can only confront a component of the system at one time. Any "break" into the sequence is an intervention with the component that is active (or, more correctly, reactive) at that particular point in time and not an intervention with the whole of the system. It is true that this componential intervention can eventually "reverberate" across the entire system (as other components react across time), but the intervention qua intervention is never considered to affect the whole system at the time in which it is performed.

As Nichols puts it, "the limitation of the communication model is that it isolates the sequence of behavior that maintains the symptoms, and focuses on two-person interactions without considering third persons."[49] This is the same problem that Bertalanffy described: Mechanistic perspectives deal only with "two-variable problems," and even these occur across time. Interventions are formulated in terms of family member A causing family member B, and B, later in time, causing family member C, and C, still later, causing family member *A* once again. This approach is two-variables *one at a time*, rather than three variables (*A, B,* and *C*) interacting *at the same time*. The circular depiction of this interaction obscures the actual linearity of this conception (see Figure 8.1).

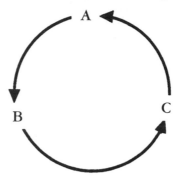

Figure 8.1   Cybernetic conception of the interaction of three family members

Figure 8.2 shows this conception as it would occur across the line of time, making its inherent linearity more apparent.

Other mechanistic (and thus Newtonian properties) are evidenced in the explanations and interventions of communication therapists. All three approaches, for example, use the concept of homeostasis.[50] This is the notion that families maintain a state of balance or equilibrium in their systems. Bertalanffy specifically tagged this conception as mechanistic in his original writings. He felt that homeostasis downplayed the active, spontaneous characteristics of any living system. He endorsed Buhler's conclusion that a contrasting organismic approach results in "a complete revision of the original homeostasis principle which emphasized exclusively the tendency toward equilibrium."[51]

The fact that homeostasis is Newtonian is clear by its emphasis upon stability and universality. Instead of the discontinuous and spontaneous change of organismic accounts, families are considered to move into homeostatic balances that leave their systems in relative stability across time. In fact, the homeostasis of a system can possess too much stability from the viewpoint of many

... B ⟶ C ⟶ A ⟶ B ...

Figure 8.2   Cybernetic conception across linear time

communication therapists, and constrain the system with rigidity and inflexibility. Similar to any mechanism, the family can become "stuck" in its current "hardware" conditions. This prevents healthy adaptation to the changing circumstances of the family. Therefore, and again in consonance with its machine counterparts, the family requires an active agent "outside" the system—a therapist in this case—to "unstick" it.

Despite these clear mechanistic propensities in theorizing, all three models—the MRI, strategic, and Milan approaches—reveal "organismic" influences in practice. Their mechanistic "two variable" theorizing sometimes yields to treatment techniques that assume "multivariable" and simultaneous interactions. Perhaps the best example is the use of paradox in treatment. The symptoms of a family problem may be "prescribed" to get the family to do what they are told *not* to do. Although the rationale of paradox is beyond the scope of this book, it is generally considered consistent with cybernetic foundations.[52]

The *practice* of paradoxical intervention, however, seems to stress taking all members of the family into account simultaneously. Indeed, some paradoxical interventions appear to address at *one* point in time the system as it is thought to occur *across* time. The Milan group makes this explicit. When the paradoxical "directive" (as it is called in the strategic approach) is given, it is expressed toward an audience and a system considered to be present simultaneously. Although the explanation for its effectiveness is often described in circular terms—e.g., first it affects mom, which then affects dad and so on—its *implementation* as an intervention seems directed toward the family as a whole. Other interventions (e.g., reframing, neutrality, and pretend techniques) are sometimes similarly practiced. Even the "circular questioning" technique of the Milan team—supposedly undergirded completely by circular causal *theory*[53]—often focuses attention upon simultaneous (as opposed to sequential) interactions in *practice*.

How is this possible? The actual mode of intervention is sometimes at variance with the explanations and theoretical assumptions that are thought to guide these interventions. How can such practices result from such theorizing? First, the aim of the family therapy movement is the treatment of the whole family. In other words, the very foundation of the field is, in some sense, nonlinear—despite the pervasiveness of cybernetic metaphors. Second, portions of communication theory (apart from cybernetics) can lead to nonlinear practices. For example, the simultaneous quality of fam-

ily interaction is implicitly acknowledged in the dictum of communication approaches: one cannot *not* communicate.[54]

In this sense, family members are not passing information to
one another in a circular causal sequence (as cybernetics would
imply). Family members are communicating to one another *constantly*, both verbally and nonverbally. Indeed, their communications can be said to form a whole at any one point in time, with this
synchrony (or symphony?) of communication *implicitly* guiding many
family therapy practices. Nonetheless, no cybernetic metaphor or
property of circular causation implies this synchronous quality of
systems. No characteristic of a mechanistic system even hints that
therapists should treat the entire system at once. Therefore, these
nonlinear aspects of communication approaches must stem from
organismic assumptions—assumptions that are not acknowledged
in theory but implicitly acknowledged in practice.

Organismic Systems Therapy—Linear Anomaly

Are there family therapies that acknowledge such organismic techniques in theory as well as practice? Although the connection to
Bertalanffy's organismic approach is rarely made explicit, there *is*
theorizing that reflects his organismic framework. Models that are
considered more structural and experiential are cases in point. These
models are also more generally anomalous to the Newtonian paradigm than communication approaches in *practice*. As in individual
and group therapies, therapists who emphasize holistic patterns
and atheoretical experiencing usually rely less upon the linear past.
Family structure is typically viewed as a nonreducible whole best
understood from its present relationship among the parts, and family
experience is typically seen as something relative to a holistic context, happening only in the present. This also results in a mini-
mization of universality and continuity which is often the reason
experiential approaches are viewed as atheoretical. These approaches deny that temporal contexts are universal; therapeutic
strategies must vary with the variance of families and their unique
situations.

Consider the well-known *structural* approach of Salvador
Minuchin. Minuchin takes many pains to avoid reductionism in his
theorizing, even reductionism across time. For instance, he bemoans
the problems of Western language in attempting to discuss relationships and wholes. He notes that the mental health field has no

term to represent even the rich "two-person unit" of mother and child: "One could coin a term, such as *mochild* or *chother*, but it would be impossible to devise terms for all the multiple units."[55] Clearly, Minuchin's intent here is to capture the "betweenness," the combination in simultaneity, of the two people. Western thinking and language tend to separate the two entities—semantically, spatially, and chronologically—and thus disregard the synchronous relation he is attempting to capture.

He proposes instead the word "holon." Coined originally by Arthur Koestler, holon comes from the Greek term *holos* ("whole") with the suffix *on* suggesting a particle or part.[56] In the organismic tradition of Bertalanffy, holon implies that the part and whole are intimately and inextricably related. Every holon is both a part and a whole simultaneously. Hence, a holon can be an individual, nuclear family, extended family, community, and so on. All these entities are both wholes and parts that are themselves contained in wholes, simultaneously rather than sequentially. Although wholes exist and change across time, their parts are not separated by time, according to Minuchin: "Part and whole contain each other in a continuing, current, and ongoing process of communication and interrelationship."[57] As applied to familial wholes, this means that family members have interrelations that are continuing, current, and ongoing. Their interaction is thus "multivariable" at each moment (in the organismic tradition), rather than "two-variable" one at a time (in the mechanistic tradition).

The term "holon," Minuchin contends, "is particularly valuable for family therapy, because *the unit of intervention* is always a holon."[58] This particular statement clearly betrays Minuchin's organismic assumptions. Instead of the unit of intervention being a particular individual or dyad of cause and effect across time, the unit of intervention is always the *structure* among the parts at the same time. Although an individual holon is a legitimate subject of therapeutic inquiry, Minuchin clearly sees *family* therapy as focusing upon the synchronous relations *among* the individual parts involved. He warns us that the conception of a holon is especially difficult for anyone brought up in Western culture. Students of family therapy are likely to overlook the relationships they are intending to focus upon. "The student may therefore have to focus rather conscientiously on the realities of interdependence and the workings of complementarity."[59]

Other aspects of Minuchin's theorizing and therapy reflect organismic assumptions. For instance, his pictorial representations

M   F
— — — — —

Children

Figure 8.3   An example of Minuchin's diagrams without linear
time

of family systems do not contain depictions of time, sequentiality,
or causality. Whereas communication approaches evidence an ever-
present "arrow" of time in their pictorializations of the system,[60]
Minuchin's diagrams depict holons in static relation to other holons
(e.g., Figure 8.3 shows a parent holon with a child holon). Minuchin
does not feel that time is irrelevant to his theorizing. Indeed, he is
most cognizant of "life cycles" and the "family over time." From his
perspective, though, holonic relations themselves do not occur across
time. That is, the structure is ever-present at any point in time and
thus can be represented by an arrowless drawing, since the passage of time is irrelevant at this level. This does not mean, however, that the structure as a simultaneous whole does not change
and differentiate.

Because all parts interact continuously, the present is as important as the past in understanding the nature of the system. The
present state of the system is not a product of the previous event in
a "reverberating chain," as in communications approaches. Past
events (as experienced by the system-as-a-whole) are important
(e.g., family crises), but there is a system existing in the present
that is worth understanding *as a whole*. An understanding of the
system's parts as they occur across time does not "add up" to the
system's parts as they occur together at any one time. Consequently,
Minuchin employs many here-and-now assessment and intervention techniques. Two of the most important are "enactments" and
"spontaneous behavioral sequences," both of which are live and
immediate demonstrations of the family structure as it *currently*
exists.

As with all family therapies, Minuchin's model has organismic assumptions mixed with mechanistic assumptions. Despite his
greater reliance upon organismic conceptions, Minuchin reflects a
few mechanistic ideas in his theorizing and therapy. For example,
he tends to employ homeostasis and equilibrium notions. He does
affirm the spontaneity, discontinuity, and open-endedness of living
systems in the tradition of Bertalanffy.[61] Still, he also agrees with

many other family therapists that families can become homeostatic, especially dysfunctional families. In addition, Minuchin clearly advances a universalist theory of family work. He often has an "ideal family" in mind, devoid of temporal specificity or context. According to Minuchin, certain universal structures of families are associated with certain kinds of problems, and hypotheses regarding dysfunctions can be developed even before the first interview. On the other hand, Minuchin warns against sweeping a priori assumptions about the family. The therapist must not hold to rigid structures and should revise hypotheses regarding such structures in light of changing systemic patterns.

This openness to the changing temporal context of the family is taken to another level in the *experiential* approach. Many experientialists have deliberately refused to create a systematic theory of psychotherapy. They wish to be open to the experiences of the family and themselves as therapists so that the process of therapy unfolds in an authentic and genuinely responsive manner. Perhaps the best example of this is Carl Whitaker—the "dean of experiential family therapists."[62] Whitaker plainly distrusts all theories: "My theory is that all theories are bad."[63] Although his language at times betrays his psychodynamic training, he feels that theory cannot encompass the varied temporal contexts in which families find themselves. In fact, Whitaker would agree with another experiential family therapist, Walter Kempler, that experiential psychotherapy has no techniques, only people and families.[64]

This atheoretical orientation seems to be a clear avoidance of the universal theorizing associated with Newtonian explanations. Indeed, this approach contrasts with most other organismic approaches. Recall that Minuchin held established notions about the ideal family and the best techniques to facilitate this ideal. Even the father of organismic systems theory, Bertalanffy, held a universalized notion of systems. He postulated that all systems had certain properties in common, hence, his notion of *general* systems theory. In this regard experientialists such as Whitaker depart from the organismic and structural tradition. Their intellectual descendancy appears to stem more from the writings of existentialists, particularly Heidegger.[65] Heidegger is a philosopher who partly made his influence known through the psychotherapy of Ludwig Binswanger and Medard Boss.[66] Heidegger (and particularly Boss) advocated nonuniversalized views of time and being.[67]

Another non-Newtonian quality of experiential family therapy is its emphasis upon the freedom of the family, both as an organization and as a set of individuals. Unlike most other approaches to family therapy, the individual or family is not considered a product of the past. There is no necessary connection to events of the linear past. If anything, people are more pulled toward the future, as represented in present images, anticipations, and goals.[68] Such images and goals are not produced by the past but rather produce the meaning of the past. Dysfunctional families are thought to be lost in their present meanings of the past. They often avoid the risk-taking and creativity required to move to a new and more functional set of goals. They have lost touch with their own power and freedom to transcend their past and effect a new future for themselves.

With this notion of families, it is little wonder that the experientialist is present-oriented in treatment. Here again, Whitaker and his colleagues are anomalies to the Newtonian paradigm and consistent, once again, with organismic assumptions. Instead of the usual interest in the familial past (e.g., previous information exchanges, transgenerational influences), experientialists focus upon the present as the most important perspective on the family. Experientialists typically avoid history-taking.[69] Information about family of origin (previous generations) may be given some attention, but it is not considered causal to the family dysfunction in the present. Therapeutic methods are also here-and-now. They are generally spontaneous and creative, themselves not dependent upon what has occurred before, including what has occurred in a recent session of therapy.

Experiential therapists are also holists in the organismic tradition. Although they attend to the needs and growth issues of individuals, they view the family as a unit. As Nichols puts it: "Whitaker sees the family as an integrated whole, not a confederation of separate individuals."[70] Indeed, for Whitaker the whole is more than the nuclear family. The whole is really the extended family (in the present), meaning three generations at a minimum: "I'm tempted to say over the phone before the first visit, 'Bring three generations or don't bother to start.' "[71] Similar to Minuchin, the synchronous context of the extended family is the important issue rather than any feedback loop or mechanistic determinism among immediate family members. Indeed, it is only in the recognition of this synchronous context that people can truly experience their freedom: "Discovering that one belongs to a whole, and that the bond can not be denied, often makes possible a new freedom to belong, and of course thereby a new ability to individuate."[72]

Cybernetics of Cybernetics

This brings us to perhaps the most anomalous theoretical conception yet devised in family therapy—the cybernetics of cybernetics. As noted, many types of family *interventions* are anomalous, but few *theories* of family therapy have escaped Newtonian and mechanistic assumptions. The cybernetics of cybernetics—or second-order cybernetics, as it is sometimes called—is the closest conception within the family therapy literature to a genuine Newtonian anomaly, though it too has difficulty transcending its linear heritage. Its linear heritage is, of course, "simple" cybernetics.[73] Recall in our historical survey that simple cybernetics has been considered revolutionary in its emphasis upon a sequence of events circling back upon themselves. Indeed, this is a distinct theoretical shift from simple linear causal frameworks. As demonstrated above, however, simple cybernetics does not diverge from the linearity manifested in absolute temporal frameworks, such as that of Newton.

Second-order cybernetics, by contrast, diverges in many respects. The main reason is its consideration of the participant/ observer. Whereas in cybernetics a separate and disinterested observer (the therapist) is thought to analyze objectively the feedback pattern of a system, second-order cybernetics assumes that the observer is part of the system itself. That is, a cybernetic relationship (feedback loop) exists not only between events observed within the system but between the observer and the observed events. Any analysis of the system is therefore dependent upon the reference frame of the observer. The observer becomes part of, or a participant in, that which is observed. Everything that is going on is entirely self-referential, "whatever you see reflects your properties."[74]

Systems, then, are in the eye of the beholder. As Becvar and Becvar note, "We make distinctions based on our own frames of reference, and we punctuate reality according to these epistemological premises."[75] Or as Keeney puts it: "we should never forget that the cybernetic system we discern is a consequence of the distinctions we happen to draw."[76] The implication is that our understanding of the system is, at least in part, context specific—i.e., relative to the observer and the observed. No universals or laws in the tradition of Newton are possible because these presume objective observers, which in no way interact with the observations made. Second-order cyberneticians freely admit that even their own theoretical assertions are self-referential. They are quite comfortable with the notion that they cannot know the truth of their assertions in any absolute way. Their truth only exists as we choose to punctuate reality in this manner.

These assertions lead to an entirely different set of implications for understanding systems. Some of these implications may even involve the linear nature of simple cybernetics. It seems to follow that if observers "literally create the world,"[77] then part of the world created might be the temporal and sequential nature of systemic events. For example, the focus of study for second-order cyberneticians shifts from the analysis of input/output sequences (as in simple cybernetics) to the analysis of holistic "perturbations."[78] Perturbations involve the simultaneous relations between systems (and their parts) rather than the mechanistic flow of events to and through systems. Thus, conventional linear notions of causality do not fit: "there is no cause and effect."[79] Moreover, the observer/observed relationship is itself simultaneous: "It is self-reference, this mutualness or *simultaneity of interactions*, that gives whole systems their sense of organizational closure, or autonomy."[80]

Nonetheless, seemingly inconsistent theoretical lapses concerning time seem to mar these obvious anomalies to the Newtonian characteristics of time—viz., objectivity, linearity, continuity, universality, and reductivity. Consider, for example, the typical pictorial representation of first- and second-order cybernetics in Figure 8.4.[81] Note the time arrows through both representations (Figure 8.4a and 8.4b). Figure 8.5 reveals how the latter representation really looks across time. Although second-order cybernetics clearly relates previously unrelated aspects of the system—the observer and observed—it does not relate them as a simultaneous whole.

This would be especially noticeable if another system (e.g., another observer) were present. Presumably, the arrow of time would include this third system in sequence, and the three systems would only interact in the mechanistic "two-variable" one-at-a-time mode, as opposed to the organismic "multivariable" at-each-moment mode. No three-variable interaction would occur at any one point in time. Furthermore, all the usual deterministic notions of cause and effect would be in effect, regardless of the holistic status of the entities causing one another. As Becvar and Becvar put it, "This structure exists as a function of *previous* mutual influence/feedback/adaptation interactions."[82] This type of linear determinism is in marked contrast to an organismic conception in which structures can emerge into wholes that are discontinuous with previous interactions.

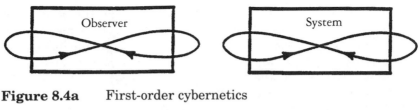

Figure 8.4a   First-order cybernetics

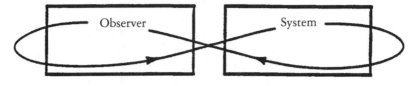

Figure 8.4b   Second-order cybernetics

Figure 8.5   Second order cybernetics across linear time

Newtonian conceptions of time tend to become even more salient as second-order cybernetic conceptions are applied to family therapy. For instance, Becvar and Becvar conceptualize the relation between therapist and family—observer and observed—in the usual action/reaction mode of Newtonian explanation: "As therapists or anyone else, we do not change systems or treat families. Rather, we change our behavior, examine the impact of this new behavior in terms of *reactions* to it, and then *react to reactions* in an ongoing modification process."[83]

It seems clear here that "interaction" is still inter-*re*action. The therapist-family system is still in a very mechanistic feedback loop. Although most other aspects of simple cybernetics are relative to the observer's self-reference frame, it appears that linear time is not one of them. Time still retains something of its absolute status because it implicitly underlies both conceptions.

Conclusion

This sojourn through the theories of family therapy shows how a subdiscipline of psychology has explicitly attempted to escape characteristics of a Newtonian temporal framework. Linear causality was identified long ago as a prime culprit in psychology's traditional reductionism to the individual. Nevertheless, the importance of time in this linearity was never recognized. This means that subsequent conceptions—conceptions intended to escape linear causality—still retain linear time implicitly. Such conceptions as cybernetics and circular causality succeed in relating all parts of the system but fail in relating them together *all at once*. Thus, the system is never a whole at any one point in time, and all the attendant properties of the whole have not been represented, studied, or applied in many family therapy techniques.

This neglect of the family "within" time has had several problematic results. Historically, the father of systems theory, Bertalanffy, has been overlooked and misinterpreted. His critique of mechanistic approaches has been lost in the literature, and his organismic alternative to mechanism has been largely ignored or linearized. Many contemporary theorists have unknowingly replaced his multivariable and simultaneous interactions with two-variable and sequential interactions. As a consequence, current therapists offer a curious mixture of linear and nonlinear techniques. Although such mixtures may be therapeutically effective, their explanations and theoretical grounds are, at best, a hodgepodge of confusing assertions about the nature of the family.

Structural and experiential approaches appear to show the least influence from Newtonian cybernetic conceptions. Structuralists value present-focused interventions into nonreducible familial wholes, without separation in time. Structuralists retain the notion that these wholes contain certain universal characteristics across time. However, experientialists seem to resist even this Newtonian property. They abdicate universal theories and continuity across time in the favor of spontaneous and unplanned interactions which are specific to the particular systemic context. A new conception—second order cybernetics—has attempted to formalize this somewhat with its participant/observer emphasis. Still, this conception succumbs to linearity in its depiction of familial events and, once again, evidences the subtlety and pervasiveness of Newton's temporal assumptions.

Chapter 9

GENERAL TEMPORAL
THEMES OF EXPLANATION

At this point, we have conducted inquiries into several of the major subdisciplines of psychology. Two questions have essentially guided each inquiry: Are Newtonian assumptions of time used widely in explanations, and are there alternatives to these explanatory assumptions? It seems clear from this investigation that each question can be answered affirmatively. That is, Newtonian assumptions of time *do* pervade each subdiscipline, and virtually every subdiscipline studied *also* manifests at least one alternate set of assumptions.

The purpose of the present chapter is to describe what these answers mean. Specifically, what does a Newtonian temporal paradigm do to and for psychology? If Kuhn's analysis of scientific paradigms is correct at all, we can expect this particular paradigm—especially given its largely unrecognized status—to have many unacknowledged effects upon the discipline.[1] For example, some theories and explanations will be favored over other theories and explanations. Some explanations may be considered "proper" or "scientific," while other explanations may find an audience with an unconscious prejudice against them. What about these "other explanations," these anomalies to the paradigm? Given their presence across the various subdisciplines, are there common themes? Do these themes point toward general alternatives to the Newtonian temporal framework for explanation?

This chapter draws upon the lessons of the previous chapters to answer these questions. We begin by distinguishing the general themes of the Newtonian paradigm, both in theory and application. Although linear time has obvious significance, its presence in theorizing cannot be fully evaluated until its implications for explanation are explicit. Of course, one cannot evaluate (or distinguish) these implications without a comparison. Consequently, the common themes of contrasting anomalies are surveyed next. These anomalies not only violate linear time assumptions, but also other concepts that have been confounded with linear time, notably causality and change. The final section of the chapter, then, describes how causality and change are conceptualized without linear time.

General Themes of the Newtonian Temporal Paradigm

The purpose of this section is to make the general themes of the Newtonian paradigm in psychology explicit. All themes are listed separately under each of the five characteristics of a linear time framework. Each theme is followed by brief illustrations of its manifestation in relevant subdisciplines as described in the previous chapters. The general question under consideration is: What are the main manifestations of linear time in psychological explanation?

The Objectivity of Time

Time is an objective standard for the measurement of psychological events. This theme pervades psychological method as well as many subdisciplines of psychology, such as development (see Chapter 2). Time is considered to be analogous to objective space. It is uniform, without content, and the bearer of events within it. It only varies in quantity or "length" and never changes its quality or nature. This, of course, makes time the perfect measure for all manner of change, and dependence upon this theme is widespread throughout psychology. Many researchers consider objective time a valuable independent or dependent measure, and many theorists have incorporated various time measurements (using clock-time as the indicator) in their explanations.

Perhaps the best example is cognitive psychology (see Chapter 5). From Ebbinghaus to the present, cognitivists have relied upon time's objective characteristics for their understanding of the human mind. First, time is thought to control, at least in part, the association of events (i.e., whether events are "close" together or contiguous in time). Second, researchers have employed reaction times as though they had a one-to-one relation with mental processes. Third, aspects of memory, such as short-term memory, are thought to have objective durations. And finally, objective methods supposedly measure rates of learning and memory—the quantity of learning in objective time.

All psychological processes occur across time. Because linear time is an objective (albeit invisible) medium for events, all psychological processes are assumed to be distributed across it (and linked by it). Relations across time are more important than relations at the same time. The former are viewed as potentially causal, whereas the latter are likely to be seen as merely coincidental. Conditioning, learning, development, personality, and experimental method are all considered to be processes that occur along the line of time. Behavior therapy is a practical extension of this, because it construes change as learning across time (see Chapter 6).

The dominance of this theme is particularly evident when psychological factors are thought to be "holistically" related. A holistic relation occurs when psychological factors are viewed as part of the same whole or process, repeatedly interacting with one another. In mainstream psychology, such factors are consistently depicted as happening across time. Indeed, the word "interaction" has become almost synonymous with sequence across time. Bandura's "reciprocal determinism" epitomizes this, because he attempts to understand the whole person through various interacting factors, yet upholds the temporal sequencing of these factors (see Chapter 3). Many therapists make the same assumption. Even many group and family therapists—perhaps the most likely to value a holistic relationship among variables (or people)—nevertheless explain their interaction in sequences across time.

Humans store the past as an objective entity. This theme notes the widespread belief that the past is an entity independent of human perception and independent of the present and future. The past is the set of moments that were once present. However, these once-present moments can be stored and recorded as any objective entity can be stored and recorded. Individual perceivers can distort the past during storage or retrieval. The term "objective" in this theme is not meant to connote accuracy but rather independence of consciousness and other temporal dimensions. Of course, the issue of accuracy itself implies that there is some objective past from which distortions can spring.

This theme is well-known in cognitive circles. Some sort of storage of the objective past is fundamental to most mainstream models of the human mind (see Chapter 5). Just as a computer stores data bits, so a human mind stores information and experiences from its past. The human mind is also capable of retrieving this past as though it were physically preserved and brought forward for remembering. Most mainstream personality and developmental formulations also require a storage capacity. Reinforcement history is an interesting construct in this regard. Although this construct is never observable (in the present), somehow its mere affiliation with the absolute past makes it seem sufficiently objective for many behaviorists to include it in their most basic explanations.

The Continuity of Time

All psychological processes are consistent with their pasts. This is one of the most powerful themes of the Newtonian temporal framework for explanation. Because time links events to one another, and time "flows" in one general direction, events of the present and future *have* to be related in some way to events of the past. This relation does not have to be causal in nature. Temporal continuity merely implies that all events that are part of the same temporal "flow" must have consistency with one another. Otherwise, they would not be a flow. In psychology, this "same flow" condition is usually met through cognitive or personality assumptions. For instance, the past is thought to be stored "in" the person so that past events have an influence on present events, regardless of whether the formal demands of causality are met (see Chapter 3). The mere supposition that past events are "in" the person gives them a cause-like influence.

Psychotherapists, of course, routinely employ this assumption. All aspects of a patient's history are potentially relevant to present behaviors and emotions because they are supposedly united "in" the patients themselves. This theme is especially strong in researchers who routinely expect consistency with the past. If their empirical findings on humans do not indicate such consistency (e.g., significant correlations), these researchers question the adequacy of their methods, not the adequacy of their temporal assumptions (see Chapter 4). Many developmentalists do not even attempt to demonstrate empirically the relevance of their work on a single age-group of people. Linear time allows them to assume that any knowledge gained about the age-group is automatically significant for (if not causal to) later age-groups (see Chapter 2).

Psychological change is continuous (if not linear) in nature. Because time is considered to "flow" uniformly (as a line), changes

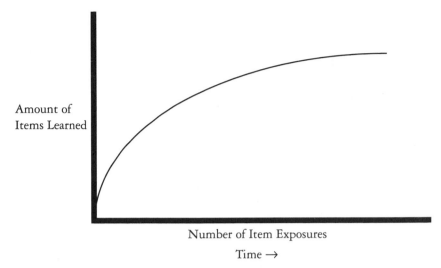

Number of Item Exposures

Time →

Figure 9.1   The assumption of temporal continuity as illustrated in a conventional learning curve.

in time always entail intervening stages and prevent any abrupt jumps into qualitatively different states. This has led psychological researchers to focus upon linear change. Psychology's emphasis upon experimental designs evidences this focus because such designs are only applicable to linear processes (see Chapter 4). Even if change is *not* linear, it is still assumed to be continuous (e.g., cyclical change). That is, the pattern of change may not be mathematically linear, but the change itself must remain gradual and quantitative in nature.

When "discontinuities" are noted between developmental stages, for example, they are likely to be viewed as faster rates of continuous change rather than abrupt qualitative changes (see Chapter 2). This is because any change is thought to require temporal duration (a line of time)—no matter how small—and thus points of time (and states of change) must occur in between developmental stages. A conventional learning curve also illustrates this (see Figure 9.1). The data of a learning curve can only occur as separated points. Figure 9.2 shows a typical array of data (whether of an individual subject or an average of subjects). Because time is considered to be continuous, intervening learning states are also *presumed* to be continuous. The data points are thus connected in a continuous flow, though no such "continuous flow" is ever observed. As a contrasting (discontinuous) explanation of these data, con-

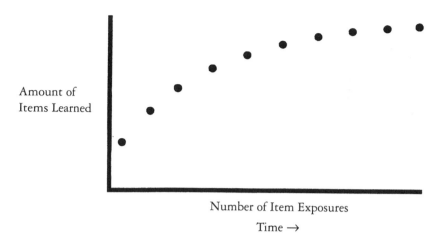

Figure 9.2   Typical array of data for conventional learning curve.

sider Figure 9.3. This figure assumes that learning occurs in "chunks" or discrete wholes, such as a whole word at a time.

The Universality of Time

Universal principles are the most fundamental forms of knowledge. True knowledge consists of universal principles that apply regardless of time (and space). Because time is uniform and continuous, any set of temporal relations (e.g., the duration of short-term memory) can be counted on to retain its original temporal pattern at any other point in time (all other conditions remaining the same). Such universal patterns are then considered "laws." Conditioning theory, for instance, is sometimes referred to as the law of effect. This law is highly prized because many view it not only as universal across all people but as universal across all animal species.[2] In fact, the more temporally universal the law is considered, the higher its status as explanation or knowledge.

This theme is also readily apparent in subdisciplines related to psychotherapy. Most diagnostic systems, for instance, assume that categories of disorder are universal across many people, i.e., basic psychopathology is not bound to a specific temporal context. From this perspective, the first step in helping a patient is finding the universal category to which they supposedly belong (e.g., schizophrenia). Then, presumably, the patient can be treated with one or more therapies associated with the category. These treatments are themselves assumed to be universal and applicable to people in the category. Some customizing to specific patient needs occurs, but the ideal envisioned by many psychotherapists is a successful matching of a universal category of disorder with a universal category of treatment.

Researchers seek conditions that facilitate the discovery of universal laws or principles.[3] Assumptions of universality place a premium on those relationships that generalize across time and tend to denigrate those relationships that do not. This means that similarities across situations are viewed as vital (e.g., stabilities and homeostatic mechanisms), whereas the unique characteristics of situations are considered "extraneous" (see Chapter 4). This theme is also related to the rationale for experimental replication. Replication is the notion that an empirical finding needs to be repeatable at different periods of time for it to be considered real or lawful (i.e., not the result of chance).

Therefore, research conditions that favor replicability and the search for similarities are the most desirable conditions. For example, laboratory experimentation and computer simulation are highly valued because they facilitate replication and exclude unique contextual differences. The assumption is that the laws discovered under such conditions govern any related context, regardless of the presumably extraneous differences found in such contexts. Cultural differences are thus considered irrelevant to, if not produced by, such basic laws.

Therapists are experts in treatment. If fundamental knowledge is a knowledge of universals, then obtaining such knowledge endows the learner with an expert status. Therapists who understand the universals of human nature are therefore experts, and should assume this role during any therapy. Although some tailoring to the specific person is usually recommended, the differences among patients are less relevant than the universal principles that supposedly lie behind these differences.

The assumption that a therapist is an expert typically results in several other temporal practices. First, universal theorizing is the hallmark of these therapies[4] (see Chapter 6). Knowledge of such theories is thought to be pivotal to most therapists' training and any therapeutic treatment. After all, it is the theorizing that contains the universals being applied. (This includes any so-called principles or laws, e.g., learning principles.) Second, treatment is typically structured to some degree. Again, the knowledge contained in universals permits therapists to know how to structure the course

of therapy so that patients garner the most benefit. Third, therapists are considered the agents of change. Because therapists are
the "experts" and patients are not, therapists need to have a firm
grip on the reins of therapy. This is the reason, for example, that
group therapists in this tradition lead "therapist-centered" groups,
with most communication occurring to and through the therapist
(see Chapter 7).

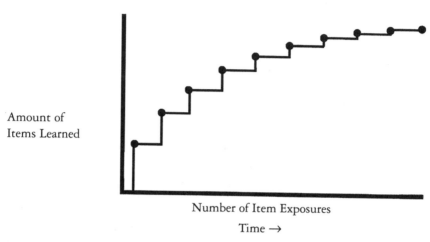

Figure 9.3   Learning data when viewed as discontinuous (or
discrete) in nature

The Linearity of Time

*Proper (or final) explanations require the antecedent determinants
of a phenomenon.* This theme is particularly reflected in psychology's
restriction to a linear form of causality. Although ahistoric, teleological, and structural forms of causality have been employed in
explaining psychological phenomena, these present-focused forms
of explanation are viewed as incomplete until "determinants" from
the past are specified. Present-oriented therapy approaches, for
example, are often criticized for their lack of "depth" or complete-
ness. Even those therapeutic approaches that focus upon the immediately preceding past (e.g., behavioral therapy) are often similarly criticized because it is assumed that crucial determinants of
the immediate past are being overlooked.

    The Bartlett question also exemplifies this theme (see Chap-
ter 5). Although cognitive constructionism implies some novel contribution by the constructor—even if only in the *re*structuring of

old memory fragments—Bartlett appeared to search for the antecedents of this novel contribution, opting for an empiricistic determination of the "construction." In this sense, "novel" contributions
by the cognizer are difficult to conceptualize—at least in the sense
of an original (and perhaps even uncaused) contribution to the
construction. Whatever is construed as "new," in this sense, is really the product of something "old"—a relay from the causal chain
stemming before the construction.

   The earlier the event, the more significant it is for explanation.
Because early events occur first in a temporal flow, they necessarily
color or set a precedent for all the events that follow. This does not
preclude the influence of later events, but it does place the greatest
weight on the earlier events for the general direction of the "flow."
Freud, despite his mixed model of personality, is thought to have
made the significance of linear temporal order abundantly clear
(see Chapter 3). The earlier the childhood event, the more fundamental it is for adult personality and behavior. Even behaviorists
affirm this general theme, though their causal chains are typically
much shorter than those of the psychoanalyst (see Chapter 6).

   Other models and subdisciplines view psychological processes
similarly. Early stages of development are often considered to lay
the foundation for later stages. In an experimental investigation,
the variable that is "before" is considered more influential (e.g.,
cause or independent variable) than the variable occurring "after."
Early cognitive "input" is thought to be stored for later mediation
of present input, with many personality theorists and individual
therapists reasoning analogously. The behavioral notion of building
blocks, for example, implicitly acknowledges that a missing "block"
of early experience is a crucial and possibly permanent deficit without some sort of remedial aid.

   Mind is a mediator between the environment and behavior.
This theme acknowledges the in-between (mediational) role of the
mind in most psychological explanations. Actually, theorists within
a Newtonian framework have no choice about this role. Virtually
any scientific phenomenon of the present has to be part of a causal
chain (or chains) that begins before its existence and continues
thereafter. The present state of the mind is no different. It is usually viewed as occurring after the environment because the existence of the environment supposedly precedes the existence of a
mind—both in physical fact and epistemological order (as favored
by empiricism). Behavior, however, follows from the environment's
influence on the mind, hence, the mind's mediational role in the
linear sequence: environment → mind → behavior.

Several historic figures in psychology are readily identified as mediationists, such as Tolman.[5] Although modern cognitivists do not subscribe to his simple S → O → R (stimulus → organism → response) terminology, they nevertheless employ models that are essentially the same (e.g., input → processing → output). The mind is still "between" and occurs after the input, whether it be input from the distant past (as memory) or input from the immediate past (as stimuli). "Hardware" conditions (innate mechanisms) are important, but they are thought to account for little of the variance of cognition. Here, "software" is considered more crucial and is repeatedly framed as the storage of past experiential "programming."

Nonlinear theorists are "linearized" or considered unscientific. Newton's framework is so influential that theorists who do not subscribe to it are often either misconstrued as linear or depicted in a pejorative manner (e.g., "unscientific" or "mystical"). Our review found several pivotal theorists treated in this manner, including Bertalanffy, Chomsky, Freud, Jung, and Kohut.

In many family therapy texts, for example, Bertalanffy's general systems theory is mistakenly assumed to be cybernetic and thus linear in temporal explanation. In some cognitive texts, Chomsky's rationalistic interactionism is reduced to innate mechanisms or linearized to empiricistic interactionism. Analogously, texts on personality theory either downplay Jung's synchronistic approach to personality or view it as a mystical (non)explanation of personality constructs. Mixtures of linear and nonlinear assumptions have fared similarly. Freud's mixed model of personality, for example, is described in many texts as purely linear in nature. Kohut's mixed approach to development and psychotherapy is also depicted in some introductory books as a unidirectional mechanistic model. In all cases, whole systems of nonlinear conceptions are either misconstrued or discounted in some manner.

Nonlinear constructs are "linearized" or construed as unscientific. This theme is derived from the fact that constructs employed by some psychologists (such as those above) do not have antecedents, by definition. The construct of free will is a good example. As used in many humanistic circles, this construct is an uncaused cause—i.e., a cause without antecedent or linear cause. Some consider a person's free will to be capable of originating a whole pattern of behaviors that cannot be determined by antecedent conditions, because this would disallow its being "free." As noted throughout the chapters, however, psychologists who employ Newtonian assumptions of time have no qualms about rejecting free will as an unscientific construct. Indeed, the construct is not conceivable from these assumptions. Antecedent events *must* determine all thoughts and behaviors without exception, though we may not yet know these determinants. Consequently, free will is rarely discussed positively in psychological texts that are "scientific" in tone.

The problem is that related constructs are sometimes used in linear explanations. The concept of choice is a case in point. Despite its frequent use in virtually all the subdisciplines of psychology, "choice" implies that the person "could be doing otherwise," and thus is not forced by any antecedents into a particular chain of events.[6] Indeed, choice often implies leaving one's particular temporal "flow," and behaving in a way that is completely inconsistent (discontinuous, nonlinear) with one's previous behaviors and attitudes. Obviously, such a construct is not compatible with a Newtonian frame, nor are a host of others, including synchronicity, ahistoric causality, group-as-a-whole, novelty, family system, teleology, equifinality, and transcendence. Nevertheless, many of these constructs continue to be used in linear explanations.

In some instances, nonlinear constructs are defined in linear terms, i.e., they are linearized. Bandura, for example, uses the term "freedom" and "transcendence" quite frequently, yet his definitions of these terms leave open the possibility of, if not require, prior determinants of these constructs. Other examples include behaviorists who have attempted to deal with humanistic objections to the lack of "free, self-directing agents" in behavioral explanatory systems. Behavioral answers to these objections typically involve the invocation of some "person variable" that is itself determined by some linear factor. Many therapists have also linearized nonlinear constructs in their formal explanations. Cybernetic explanations of family systems accomplish this, even though the father of general systems theory, Bertalanffy, contended that living systems do not have such linear properties (see Chapter 8).

The Reductivity of Time

Psychological processes are reduced to their separate parts as they occur across time. This theme concerns the widespread empirical practice of studying one particular portion of a process at a time. In fact, from a Newtonian perspective this is the only way in which a process *can* be directly studied. Because the process occurs piece-by-piece across time, it can only be directly observed one piece at a time. This is normally not perceived as reductionism in psychology because linear time is already thought to have related the pieces *as a whole*. For example, before any investigation of the pieces is conducted, we supposedly know which piece is the most important (the first) and how change occurs (continuously). Knowing these relationships a priori allows psychologists, then, to reduce their study immediately to the various pieces.

The investigations of many developmental psychologists seem to illustrate this (see Chapter 2). Some researchers study one particular age-group as if its meaning to the whole of development is obvious. Although sophisticated theories of the developmental whole have been formulated, many researchers study pieces of this whole as if they are independent units connected by the stream of time. As a part of this stream, their significance to later age-groups is considered axiomatic. Many cognitive psychologists make similar assumptions about the stream of information being processed. Information is often depicted as a linear flow of "atomic units" or "bits" separated in time. This is the reason that such information is *processed* over time. This has also caused those attempting to simulate human cognition (artificial intelligence) considerable difficulty. Although the bits can themselves be input, their independence from one another makes any information about their holistic relation difficult to encode (see Chapter 5).

Subject and object are separated in time. A repeated theme of the previous chapters is the separation of subject and object. Subjective factors are "in here" in the mind or emotionality of the person, while objective factors are "out there" in the physical aspects of the environment. Part of the modern reason for maintaining this separation is the assumed linear sequence of their relationship.[7] Like all other events or processes, subject/object processes (e.g., mind/environment relationships) are thought to be distributed across the flow of absolute time. This results in their being forever separated in time, first one and then the other, as they appear across our window of time—the present. Just as the sequencing of informational bits or developmental stages results in their being considered independent units across time, so the sequencing of subject and object results in their being considered independent units across time.

This sequentiality also implies that any interaction between subject and object must begin with one or the other (and not both). The order of this sequence happens to be a decisive factor in psychological theorizing. Explanations that are empiricist in nature assume the object precedes and thus determines (to some degree) the subject. Many behavioral, cognitive, and some psychoanalytic explanations appear to take this conceptual tack, emphasizing the environment's ultimate determination of the mind. On the other hand, many rationalistic explanations assume that the subject precedes and determines (to some degree) the object.[8] Many humanistic and some psychoanalytic theorists adopt this style of explanation, stressing the mind's organization and construction of the environment's meaning (see Chapter 3). Second-order cybernetics does not postulate any overt primacy of subject or object, yet this explanation maintains their sequence in time and therefore their conceptual separation (see Chapter 8).

The focus of psychological study and explanation is the individual. The Newtonian framework for time contributes to the individualism seen in so many subdisciplines of psychology. Historically, the appearance of linear time moved explanation from an emphasis upon present relations to an emphasis upon past relations. Instead of explaining persons in relation to other persons (in the present), persons are explained in relation to themselves as individuals in the past (e.g., developmental explanation). Individuals are now thought to contain the past (in some sort of storage) which determines them in the present. No recourse to contextual or interpersonal relations of the present is typically considered necessary, though it may be a secondary consideration.

Recent conceptions of group and family therapy have challenged this type of individualism. These therapists contend that present interpersonal relations are important to any understanding of the patients they treat. Nonetheless, many of these therapists retain another linear-based individualism. Because the group or family interaction is considered to occur across a span of time, each part of the "system" is separated *by* time. This means that the therapist experiences individual pieces of the system in sequence, with the pieces each separated in time. These pieces could be the individual members of the group (or family), or the systemic behavioral patterns (however defined). The point is that these individual pieces are thought to be experienced (and remembered) in dyadic relations only—as each piece leads to another across time—and not holistic relations as each piece relates to *all* other pieces at the *same* time. The ultimate result is a subtle, but very real, reductionism to individual parts (or relations) within the system (see Chapter 8).

General Anomalies to the Newtonian Temporal Paradigm

We turn now to anomalies in psychology. We have just delineated the dominant general themes of explanation in psychology, as related to Newton's temporal framework. The purpose of the present section is to identify any common themes of the anomalies to this pervasive framework. This not only gives a more complete (and balanced) picture of psychological explanation, as it has been reviewed here, but it also provides some contrast with paradigmatic themes. The problem with outlining the themes of an unrecognized paradigm is that there is often no contrast apparent. Distinctions and themes do not make sense, unless there is something to distinguish them from, something to compare them with. At the very least, then, contrasting anomalies facilitate a better understanding of the temporal themes now influential in psychology.

Anomalies, however, can also indicate the route to alternative explanatory frameworks. Alternative frameworks can open the way to optional explanations of the "data" as well as other "hypotheses" to consider in understanding psychological phenomena. This would also give scientists a framework against which to evaluate the Newtonian framework. Even if linear assumptions are ultimately confirmed, they will have been *evaluated*, and their influence on various topics in psychology better known. We begin the search for alternative frameworks by organizing common themes of these anomalies under the five characteristics of a Newtonian temporal framework. A summary of each theme is stated first. Then, a description follows that illustrates its use by the subdisciplines in which it is observed.

The Objectivity of Time

Time is context-bound. Anomalous themes have developed around the supposed objectivity of linear time. Rather than time being an objective entity that has certain "effects" on the mind and behavior, time is considered to be a shared (intersubjective) organization of reality. The scientist is not seen as an objective observer, but a participant/observer in which his or her own culture is paramount. In this sense, the use and perception of time depends upon the context in which it occurs. Instead of time being a contextless container of durationless moments (points on a line), time is dependent upon context (see Chapter 4). Even the controversy about daylight–saving time (see Introduction) illustrates how temporal structures can be considered reality, when they actually stem from cultural factors. The fact that several languages function without tense and several cultures have historically avoided linear notions can also be considered support for the cultural nature of time (see Chapter 1).[9]

In psychology, the influence of Kant, Heidegger, and other philosophers has led some, even in our Western culture, to question the objectivity of linear time (and the primacy of the past). Many existentialists, for example, hold that this objectivity is actually a concretization of our own organization of reality (to avoid responsibility for it). For Jung, too, the psyche "conditions" time. Several developmentalists question the automatic prioritizing of the past in explanation. And some cognitive theorists argue that the "objects" of input—even before their "reception"—are themselves constructed rather than accepted as a reality that requires "processing."

Psychological processes are viewed as "outside" linear time. Several psychological theorists have questioned the assumption that psychological processes must be viewed as occurring "in" and "across" an objective medium of time. These theorists have advanced explanations that they *explicitly* claim to be "outside" the usual linear sequence. These include Rychlak in learning and cognition, Yalom and May in psychotherapy, and Lewin and Jung in personality. (Although less explicit about their conceptions being outside linear time, other nonlinear psychological theorists include Kagan, Piaget, Rogers, Minuchin, and Whitaker.)

Perhaps most illustrative of theorizing "outside" time is Rychlak's work, which challenges the bastions of the Newtonian framework: learning and memory (see Chapter 5). Learning and memory are almost synonymous with accumulation and progress across time. Still, Rychlak advocates (and has empirically tested) notions of learning that rely more on logical rather than chronological relations. Learning is a logical (and timeless) elaboration rather than a chronological accumulation. That is, learning is an extension of already implicit meanings (or structures) rather than a store of things "taken in" from the environment. Yalom and May also question psychotherapy's traditional emphasis upon the past, as well as traditional therapeutic sequences across time, such as assessment-treatment and life-death (see Chapter 6). Such sequences are better understood as simultaneous meanings rather than objective entities separated in time.

The past is constructed from the vantage point of the present. This anomalous theme challenges the objectivity of the past. Instead of the past being an immutable set of present moments "gone by," the past is a dynamic meaning that is still influenced by and influencing the present (and future). In fact, for the hermeneuticist, past, present, and future are not three dimensions in sequence but rather three dimensions in simultaneity (see Chapter 4). If the contents of one dimension are altered, they are all altered. The past, in this sense, is not "stored" because there is no object of perception to be stored. Rather, our memories are changing amalgams of meaning, quite sensitive to what is happening in the currently lived context.

In this regard, several cognitive constructivists have noted the effects of present mood and context on memory (see Chapter 5). Interestingly, some social constructivists have even contended that developmental researchers "invented" the primacy of early development in psychology (see Chapter 2). Several linguists have also challenged the notion that meanings stem from the objective past. Saussure, for example, believes that meanings are better understood in light of simultaneous (synchronic) relations among terms, rather than sequential (diachronic) analyses of their origins. Rychlak views meanings as (teleologically) caused by the goals and purposes of the meaning-giver (in the present), without those goals and purposes being themselves determined by antecedents (see Chapter 5).

The Continuity of Time

Psychological processes can be inconsistent with their pasts. As Newton saw it, a complete knowledge of conditions at Time 1 is sufficient to predict later conditions at Time 2. Events of the same temporal "flow" must be consistent with one another. If, however, events of the same flow are inconsistent with one another, then the pattern at Time 2 is *qualitatively* different from the pattern at Time 1 and, hence, unpredictable (at least in a linear sense). It should be noted that this does not preclude discontinuous prediction. Scientists can observe qualitative changes repeatedly and predict them under certain conditions (e.g., water from hydrogen and oxygen gases). Although this type of prediction is based upon a prior knowledge of sorts, it is not based upon knowledge at Time 1 only. Complete knowledge of hydrogen and oxygen gases does not lead to a prediction of water. Hence, Time 1 and Time 2 are not continuous or consistent with one another in a linear sense, though they can be predictable in a nonlinear sense.

This lack of linear consistency is one of the ways to interpret the "discontinuous" anomalies of developmental psychology (see Chapter 2). Apparently, early developmental patterns are not always consistent with later life patterns. In fact, some aspects of development appear to be temporary adaptations that have little or nothing to do with later development. Several life-span developmentalists have concluded that humans have a capacity for discontinuous change across their entire lives. That is, humans do not have to remain consistent with their pasts. From the perspective of some theorists, persons have a free will that influences behavior and is not a product of antecedents. Many therapists also postulate such nonlinear causes. Indeed, the transcendence of previous patterns of behavior and cognition is sometimes considered the key to effective therapy. This transcendence is thought to be accomplished in many ways—from the holistic "cohesion" of a group to the "responsibility facilitation" of an individual.

Change can be discontinuous in nature. Rather than psychological change always being a smooth and gradual movement across time, change is sometimes an abrupt and cataclysmic shift in an entire pattern of events—i.e., a discontinuous change. Such a change could be seen as instantaneous in its temporal "length," and qualitatively different from its original nature.

As mentioned above, several developmental theorists (e.g., Piaget) have viewed developmental stages in this manner. Some cognitive psychologists also view mental processes discontinuously. Instead of a continuous "stream" of thought and information in cognition, some consider the mind to flit from one whole thought or perception to the next. Insight is sometimes considered to function in this manner, with the gestalt of thinking suddenly altered. Many individual therapists report seeing this type of discontinuous change in their patients as behavior and attitudes abruptly shift. In groups or families, too, a key alteration in even one person's attitude or behavior can mean a totally different group atmosphere. For this reason, Kurt Lewin considered time to be a series of momentary gestalts that remain sufficiently discrete from one another to allow discontinuous change.

The Universality of Time

"Universal" principles are specific to context. Universal laws require an objective and continuous time that flows without regard to context or culture. If linear time is not objective, however, then the intersubjectivity (shared meanings) of time could vary from culture to culture. The linear meanings of our own culture would be the result of an agreed upon interpretation of change rather than a universal truth. This would jeopardize Newton's notion that time is continuous and uniform across history. Psychological "laws"—themselves dependent to some degree on time—could vary from culture to culture and historical period to historical period.

Consider Gergen's view of social psychology in Chapter 4. For him, social relations are completely embedded within cultural norms so that no absolute laws of social psychology can exist. Social psychological findings in the United States—however fundamental and lawful they might seem—would not *necessarily* apply to social relations in Japan. Even within a culture, psychological relations can be discontinuous from historical period to historical period. According to Saussure, for example, linguistic relations and meanings at one point in time can qualitatively differ with linguistic relations and meanings at another point in time. The discontinuity of some stage theories of development also illustrates this. If stages of development differ qualitatively, then knowledge about one stage (or time period) does not necessarily reflect upon knowledge about another stage (see Chapter 2). Hermeneuticists seem to take this even further. Many of these psychologists argue that context specificity extends beyond culture to the unique situation in which research is conducted. In an important sense, therefore, knowledge is not universal because it can only be gathered in particular settings, with particular people, and so forth.

Researchers should study phenomena that are filled with context. Many anomalous researchers, such as those embracing systems methods, avoid laboratory studies and computer simulations because these research settings do not include natural context. Natural context is thought to be part of the gestalt or meaning of whatever is being studied. Even slight differences between situational contexts have the potential to alter these meanings and gestalts quite drastically. From a hermeneutical perspective, all knowledge is "situated" in the present context, and universalized principles (in the sense of cutting across *different* temporal contexts) are not meaningful forms of knowledge (see Chapter 4).[10]

The experimental search for such universals is analogous to examining the similarities among qualitatively different developmental stages (see Chapter 2). Although knowledge of the aggregate of these stages is perhaps interesting, it tells us little about the nature of any particular stage. Indeed, no application of an aggregate "law of stages" could ever provide an understanding of a particular stage's distinct qualities (or a child existing in that stage). Analogously, Dreyfus claims that no application of universal cognitive rules (as contained in artificial intelligence devices) could ever provide an understanding of everyday behaving (see Chapter 5). The very nature of such rules—cutting across the uniquenesses of situations—precludes their applicability to specific, context-laden environments. Universals are applicable only when context across situations is similar, i.e., when changes in context do not alter the gestalt of the process under consideration.

Therapists are not "experts" in treatment. Because therapy cases are potentially different in quality, each case must be understood in relation to that quality. That is, each case is temporally "situated" within the specific context of the case rather than part of a consistent "flow" across contexts (or people). Rules of thumb can be helpful guides, but they are ultimately expected to be—at some level or other—inapplicable to the person or persons requiring treatment. Each person's unique mix of life factors means a different—and possibly discontinuous—gestalt of therapeutic issues. Consequently, therapists have no knowledge that would permit them to be experts in this sense.

Nor can a therapist become too reliant on universalized theories.[11] As noted in Chapters 6 through 8, many psychotherapists—including Jung, Whitaker, May, Yalom, and Minuchin—caution against relying on theory for fear of missing the person or persons as they *really* (uniquely) are. The therapist must be open to the patient. If the therapist imposes his or her own structure (from some universalist notion of therapy), then the unique gestalt of the particular patient could be overlooked. Anomalous therapists typically give the patients more control of their treatment than is typical of most mainstream therapists. Indeed, most anomalous therapists view the patients, rather than the therapists, as the agents of therapeutic change.

The Linearity of Time

The present can determine psychological phenomena. Although the past is considered an important (nonobjective) context for the present, determinants of an event or process can be simultaneous (or "synchronous" or "contemporaneous"). The simultaneous conditions of the present are integrated in such a fashion that a new gestalt emerges. This gestalt is not caused by events of the immediate or distant past and is instead the result of the present relationship of its parts. The present quality of each part is, in turn, determined by its relation to the present whole. In this sense, the relevance of an event to present meanings and patterns better indicates its significance than its linear order in time.

The practical manifestations of this anomalous theme are seen in present-focused psychotherapies. These treatments postulate that all the necessary conditions for assessment and treatment exist in the present. Further, the indicator of therapeutic importance is relevance to the now rather than chronological primacy of the then. Some object relations approaches take this tack at times, as well as here-and-now group therapies and structural family therapies. The present cognitive construction of the past is another example of this anomalous theme. From a rationalistic constructionist's perspective, the past is no longer causal in the traditional linear sense (see Chapter 5). The past takes on a role more akin to "ground" in a figure/ground relationship. The past does not determine the present meaning but provides a simultaneous (present) background that contributes to its overall gestalt. Saussure claims that present conditions determine language in this manner. The roots of meanings are not discerned from historical linguistic changes; word meaning is understood from its present relation to other word meanings.

The "future" can determine psychological phenomena. The future here is not a linear future in the sense of occurring after the present (hence, the quotation marks). The future here pertains to meanings, goals, and purposes existing in the now. Teleologists (or final causal theorists) who emphasize these as causes or determinants do not view them as being determined by linear causes.[12] They view final causes as the initiators of actions and attitudes, independent of antecedent determinants. This is not to say that such initiators are independent of context which may include the "past" (in the present). Indeed, teleologists typically contend that a creative (or constructive) past is vital information in the formulation of goals and purposes. Nevertheless, this information never *determines* such formulations.

Several theorists of the previous chapters are supportive of this more telic view of human nature. Bertalanffy, for instance, holds that living systems evidence a kind of future orientation he calls "equifinality" (see Chapter 4). Virtually all the psychodynamic theorists reviewed—Freud, Jung, and Adler—proffer some telic aspect in their theorizing. Although these aspects are often "linearized" or downplayed by modern interpreters, they are nonetheless quite prominent (see Chapter 3). In the cases of Adler and Jung, teleology is one of their most explicit and important assumptions. Even Adler's conception of "motion" is thoroughly telic in meaning (see Chapter 6). Many existentialists also evidence this telic flavor in their theorizing. Yalom, for example, states that the most important tense is the "future becoming present," rather than the present becoming past. Even the cognitive theorist, Rychlak, feels that he has telic explanations for memory *and* empirical demonstrations to support them (see Chapter 5).

The mind and environment are simultaneous parts of a holistic context. If linear time is not held to be absolute, then events of the world do not have to operate in absolute sequences. This permits many psychological processes to be interpreted differently. The mind and its environment, for example, can be framed as simultaneous and holistic rather than sequential and mediational. From a Newtonian perspective, the environment is separated from the mind, and must first cross this separation (across time) to reach and affect the mind. Once some representation of the environment is "in" the mind, it processes this representation in a manner that is reflected to some degree in the person's behavior (see Chapter 5).

In our review, Lewin probably spoke most directly to this issue. For him, cognition and the world are simultaneous parts of a greater gestalt called the life-space. Because the two parts are already *in* the life-space, they do not need to cross time (and space) to reach each other. The world does not have to be transmitted to the person in the manner of "input," nor does the mind extend messages and behaviors to the world in the manner of "output." Lewin's conception obviates any "in between" or mediational role for the mind. Variations upon this gestalt theme include Jung's synchronistic notion of mind, Rychlak's telic approach to cognition, and Dreyfus's description of human action.

Humans are agents of their own actions and attitudes. This anomalous theme stems from several types of nonlinear explanations and constructs, including "free will," "self-transcendence," "agency," and "volition." A common theme among these explanations is that humans are responsible for their own actions. This is considered impossible in a Newtonian framework, at least in the sense of the person *originating* any pattern of behaviors. From a Newtonian perspective, the past is the agent of any person's actions. The person may be responsible for behaviors in the sense of having stored information that is triggered in the present, but the person cannot make an original contribution that is not already determined by antecedents.

Many "third-force" theorists and therapists have challenged this perspective. They feel that some escape from "past determinants" is vital to responsibility and human dignity. Even teleologists who see humans as being determined by future goals and meanings feel there must be an "agent" (the person) formulating these goals and meanings (see Chapter 5). Many existential and some psychodynamic approaches to therapy also reveal this assumption. Patients are encouraged to assume responsibility for their own actions. Although such patients can, of course, be victims of someone else's actions (e.g., physical abuse), someone else's actions are not viewed as the determinant of a patient's meanings. From this perspective, there is always hope of transcending such traumas and reconstructing the meanings of the past.

The Reductivity of Time

Psychological processes are viewed holistically. This anomalous theme asserts that individual parts of a process cannot be fully understood without reference to the whole. Just because parts seem to occur separately in time does not mean they are not influenced by the whole of the process. Moreover, properties of a whole process are not adequately described by their supposed linkage "in" linear time. For instance, a linear process would preclude a later event affecting an earlier event. From a holistic perspective, however, the later event may be part of a complete whole that lends meaning to the earlier event (e.g., last words of a sentence). Indeed, the "final" event may be pivotal in that its contribution changes the entire gestalt (and hence, identity) of the "preceding" events (as taken together).[13]

Many family therapists appear to confirm this holism. They see the system of a family as having a holistic meaning—a gestalt of all parts—even though the system may be realized only in parts across time. In this sense, the meaning of earlier parts cannot be known until later parts manifest themselves. The same is true for some developmental theorists. They admit that they do not know the part played by a developmental stage until they have some sense of the developmental process as a whole. Likewise, Adler's "law of motion"—though seemingly consisting of parts across time— is nevertheless united by the telic relationship among its parts (see Chapter 6). Dreyfus makes a similar point regarding artificial intelligence (AI) machines. Because such machines process information one piece at a time, they cannot discern the holistic meaning of each part (see Chapter 5). Such meaning can only come from the information-as-a-whole, all bits perceived simultaneously. AI machines cannot attempt this "whole" until they have received all the information bits, and even then, there is no information about the bits' original (simultaneous) arrangement. Humans, according to Dreyfus, naturally see and understand the holistic qualities of parts.

Subject and object are understood as a simultaneous whole. A major theme of a Newtonian- and Lockean-based epistemology is its temporal separation of subject and object. If all processes occur across time, then processes involving subject and object also occur across time, one preceding the other. The only issue remaining is "which comes first—subject or object?" Psychology's empiricistic roots generally give the nod to the object (i.e., the environment). Because the environment precedes the existence of the person, and the person is viewed as relatively inactive at birth (in the tradition of the Lockean tabula rasa), the object is thought to precede the subject. On the other hand, a small minority of theorists tend to favor the subject first. The mind is viewed as structuring the world before any object is ever "processed."

A number of anomalous theorists, however, dispute both sequences. Those who challenge the notion that the mind is a mediator would certainly qualify here (e.g., Lewin and Rychlak). In addition, some systems theorists, such as Bertalanffy, hint at a subject/object integration. Bertalanffy's notion of an organismic system—a nonreducible and simultaneous whole—seems to assume this. Heidegger, Dreyfus, Piaget, Yalom, and May all make the simultaneity and integration of subject and object *explicit*. The existentialists of our review, for the most part, follow Heidegger. He consolidates subject and object in his notion of *Dasein*.[14] It is also no coincidence that *Dasein* merges the three dimensions of time—past, present, and future—preventing any sequentiality of subject and object, and their consequent separation in time (see Chapter 4).

Individuals are understood in relation to their simultaneous contexts. Individualism is a hallmark of the Newtonian legacy for psychology. Individuals supposedly house their pasts and are separated in time from other individuals as they interact in sequence. Anomalous explanations, however, dispute this type of individualism in both respects. First, the past is not stored in the individual but is instead part of the present context and culture. In this sense, important qualities of the individual do not inhere in the individual but stem from the individual's relation to the world. Second, individuals are not separable in time because relationships do not occur in sequences. Although speaking one at a time is a social custom, individuals are nevertheless *always* communicating with one another, whether verbally or nonverbally. Individuals are thus part of a greater synchronous whole, termed variously as *Dasein* or system, which supersedes psychology's traditional separation of individuals in time.[15]

The family therapist, Minuchin, seems to argue for this with his concept of a holon (see Chapter 8). Holons recognize the continual (and simultaneous nature) of ongoing relationships. At the same time that a mother is speaking, her child is communicating nonverbally. Minuchin argues that it is wrong to see the mother and child in sequence. At the very least, sequencing omits half of the simultaneous whole, because attention is paid only to one member of the whole at each time point. We should recognize, according to Minuchin, that the two individuals are better conceptualized as two aspects of one entity—a holonic "mochild" or "chother." Other anomalous explanations make similar contentions, including some object relations theories, group-as-a-whole approaches, and second-order cybernetics. Second-order cybernetics clearly aspires to this holonic status. It attempts to relate the observed, the observer, and the accompanying context. Unfortunately, this explanation falls back on its mechanistic roots when it sequences these factors and separates them in time.

Implications of the Anomalies for Causation and Change

The common themes of these anomalies stand in stark contrast to the general themes of a Newtonian temporal framework. Indeed, anomalous themes challenge virtually every aspect of this framework, from time's objectivity to its reductivity. What about other concepts long associated with linear time? As noted throughout the book, many of the most important concepts in the discipline, such as causality, change, process, and even behavior, involve absolute time. Do we have to abandon these concepts when anomalies to linear time are considered?

Clearly we do not. In fact, the only reason this question arises is that these concepts have been inappropriately confounded with linear time. Some of the difficulties in explaining psychological phenomena stem from this confounding. Definitions of concepts such as causation have been so restricted that alternate definitions have been outlawed (as unscientific or mystical), without any formal investigation of their appropriateness. The purpose of this section, then, is to begin the disentangling of linear time from pivotal explanatory concepts, particularly causality and change, so that the suitability of other definitions can be evaluated.

Separating Causality and Linear Time

Let's begin with perhaps the most crucial concept of scientific explanation—causality. Causality has long been construed as occurring in sequence across time, cause always preceding effect. Because time flows, all processes—including causal ones—must take place in sequence across this temporal frame of reference. Several empiricists, notably Hume, have upheld the temporal antecedence of cause over effect.[16] The combination of Newton's metaphysics and Hume's epistemology has led most psychologists to view all psychological phenomena as determined by some type of linear chain (or chains) of causes and effects. Linear causation has also provided psychology with its main objective—discovering those causes that precede the effects of interest. This accounts for the almost exclusive focus of research and therapy upon the past.

The question here is: Does causality require linear time? Is temporal antecedence necessary to conceptualize causality? Perhaps surprisingly, analyses of causality have *not* supported this assumption. In his pivotal books on causality, for example, the physicist and philosopher Mario Bunge claims that "the principle of antecedence and the causal principle are independent of each other."[17] After reviewing prominent formulations of the causal principle, he finds that "none of them contains the idea of temporal priority."[18] Nonetheless, as Bunge notes:

> The confusion between antecedence and causation is so common that philosophers have found it necessary, very long ago, to coin a special phrase to brand this fallacy, namely, post hoc, ergo propter hoc (after that, hence because of that). It is not physically possible for the effect to precede its cause, since the cause is supposed to give rise to, or to produce, or to contribute to the production of the effect, e.g., by means of an energy transfer. But it would be logically possible for the cause and the effect to be *simultaneous* in a given frame of reference, particularly if they occurred at the same place. Even if the cause and the effect took place at different regions in space, one could imagine that a physical agent might traverse a distance in no time—as has been held by most theories of action at a distance. And such an *instantaneous* action at a distance would be perfectly compatible with causality—Hume and Humeans notwithstanding.[19]

Bunge makes clear that a logical (or existential) priority of cause is necessary because the cause must be present for the effect to occur. In our linear world, however, this *logical* priority has been interpreted to mean *chronological* priority. That is, the fact that the cause needs to be present (as opposed to absent) has come to mean that the cause needs to be before (as opposed to after). Causality, though, is quite compatible with what Bunge calls "instantaneous links."[20] Cause and effect can be simultaneous. Indeed, cause and effect do not even have to be contiguous in space. According to Bunge, a cause-effect relation can be instantaneous, even if the events of the relation are vastly separated in space.

The problem in psychology is that researchers view simultaneous connections as an indication that events are *not* causally related. Jung, for instance, noted many important synchronistic events that are simultaneous (and separated by space). Such synchronistic relations are generally viewed as unscientific and noncausal (see Chapter 3). This is due, at least in part, to the simultaneous occurrence of these relations, precluding any transmission of force between the events. If, however, temporal antecedence is not necessary to causal relation, then it is possible that a multitude of psychologically significant (synchronous) variables are being overlooked merely because one variable is not observed to precede the other in time.

This possibility is seen most readily in those philosophers and psychologists who contend that linear causation is just one of several forms of causation available for theorizing—the remainder of which are rarely used in psychology.[21,22] As far back as Aristotle, scholars have used no less than four general categories of causality. The only one really compatible with Newtonian formulations of time was called "efficient causation" because it required movement (*efficiens*) across time.[23] With the historic rise of linear time, the "four causes" were narrowed to one—efficient causation—at least by the lay culture.[24] Antecedence became a requirement of causation, and all psychological processes are now presumed to be the effects of linear chains across absolute time.

The other three "causes"—final, formal, and material—do not require absolute or linear assumptions of time. These three establish their effects through instantaneous connections rather than the sequential links necessitated by efficient causality. Final cause (or teleology) establishes its effect through the purpose something has. Although this purpose can be viewed as a goal in the linear future,[25] many teleologists consider purpose to be simultaneous with the present, a present image of the "future."[26] Formal and material causations have been previously alluded to in structural terms. Both emphasize the static and timeless organization of something— matter being the emphasis of the material cause, and pure form (e.g., "gestalt" or "essence") the emphasis of the formal cause.

Needless to say, many of the anomalies described in the previous chapters are explainable through these alternate causal analyses. Bertalanffy's notion that open systems, such as families, are directed toward some future goal is understandable in terms of final causation. This also meshes nicely with his contention that organismic systems function synchronously.[27] Formal causal theorizing is illustrated in Piaget's notion of subject/object interactions. He saw such interactions as simultaneous gestalt patterns rather than sequential chains of efficient causality.[28] Indeed, Piaget's notion of cognitive biological structure is an example of a material causal explanation (see Chapter 5). Such structures are thought to aid in the "construction" of reality when the newborn is first "acting on the world" (e.g., sucking reflex). Neither the "constructing" nor the "acting on," however, is considered to occur in an efficient causal manner.[29]

Of course, the narrowing of theoretical options (and causal assumptions) is the scientist's prerogative. It is perhaps important and even necessary, at some point, to narrow the choice of explanations in order to understand best the particular phenomenon at hand. The point here is that the current narrowing of causation in psychology has occurred because of its relationship to an unrecognized assumption (linear time), *not* because it is necessarily the best one for the phenomena under study. Efficient causation may indeed be the best explanation, but until other options are evaluated, this cannot be known. The problem is that these other options have been ruled out without investigation because they do not fit a particular view of time which itself has been assumed without evaluation.

Some psychologists may contend that a necessary narrowing of causal options has occurred through the evolution of the natural sciences. That is, Newton and other scientists have evaluated the various forms of causal assumption over the years and found all but efficient causation deficient. The problem with this contention, however, is twofold. First, this supposes that some sort of semiformal or quasi-scientific evaluation of these causal assumptions was undertaken, when there is no evidence of this. In Newton's case, for instance, he uncritically accepted his assumptions of causality from his predecessors.[30] Second, this contention assumes that the natural sciences do not currently employ the other three causes, when this appears to be an entirely false assumption. Bunge, for example, holds that biology and physics employ *all four forms* of explanation.[31]

The crucial issue is the precipitous narrowing of psychology's explanatory options. Because of an unrecognized and unevaluated assumption (linear time), psychology's explanations have been largely restricted to the efficient causal variety. It is therefore vital that we formally identify manifestations of efficient causation in

psychology to become more aware of this restriction. Causal restriction is not, of course, the only form of linear limitation on
psychology. Psychology has also unknowingly limited its explanations of *change*. Change, like causality, is currently confounded
with linear time in psychology. Therefore, it is important that we
attempt to differentiate change from linear time before proceeding
to alternative notions of time in the final chapter.

Separating Change and Linear Time

The concept of change seems to have met the same historic fate as
that of causality: fusion with linear time. Because time is uniform
("flowing equably"), change is also expected to occur in a smooth
and orderly fashion, much like a flower blooms. Change must be
consistent with its past (like time), allowing only gradual (and
predictable) movement from immediately preceding events. This
means that change has temporal duration and sequence. Change
occurs across some interval of time and must be tracked through
some sequence of observations. There must be a minimum of one
observation before the change occurs and one observation after the
change process has ended—no matter how small this interval of
time may be. Change, then, can be measured objectively as the
difference between two or more positions on the temporal continuum.

As it happens, however, this is only one conception of change—
sometimes called *quantitative* change to distinguish it from *qualitative* change.[32] Similar to its effect on causality, the assumption of
absolute time has restricted the conceptualizations available for
understanding change. Quantitative change assumes that change
is a modification in the "quantity" of something—its size, volume,
or length. Because Newton viewed time as a grand number line (a
continuum), time and change both have a quantifiable "length."
Some "points of time" are always *in between* the beginning and
ending of any change process. Change can only occur through smooth
and incremental increases in its quantity. Accelerated growth or
change is permitted (a shorter "length" of time), but this still assumes that the growth entails at least *some* points of time, and
thus some states of growth in between beginning and ending points.

Qualitative change, by contrast, refers to the sudden alteration of an entity's basic nature or quality. Qualitative change is
not typically permitted in a Newtonian frame of reference because
no passage of time (or sequence of observations) is necessary for it
to occur. It occurs, rather, in *instantaneous* shifts between states or

qualities—without movement through the gradations of time and space supposedly needed between the states (as Bunge noted above). Shifts, in this sense, are abrupt or discontinuous. If qualitative shifts occurred smoothly—or by small degrees—then they would be better described as changes in *quantity* (of the same thing) rather than changes in *quality* (to a different thing). For this reason, holistic change is usually considered qualitative. Although only one small part of a whole may change, its "effect" upon the whole can be cataclysmic, altering the entire meaning (or relationship) of the other parts instantaneously.

This also implies that qualitative change can be observed all at once. Unlike quantitative change, it does not require two or more observations to track the beginning and ending of a change process, because the change "process" is instantaneous (no temporal "length"). Before and after observations can, of course, pick up qualitative changes (if the observations are positioned accurately).[33] Indeed, there *is* a type of sequencing involved in such change, as the entity discontinuously shifts from one quality to another. Still, this particular type of sequence is not linear in nature. This sequence can be viewed as temporal (see Chapter 10), but it definitely lacks the characteristics of time espoused by Newton. It does not involve a period or "length" of time, nor does it require any continuity (or consistency) from one moment to the next. In this nonlinear sense, the designation of "before" and "after" merely implies the order of the differences observed.

Consider Piaget's theory of cognitive development in this regard. Piaget asserted that all children go through the same stages in the same sequence.[34] Nonetheless, the reason for this sequence is not because the stages are continuous with one another in the linear sense of one billiard ball hitting the other across time. Indeed, the stages are already implicit (all at once) in each child's genetic code. Piaget recognized that each stage would not *be* a stage without its discontinuity from the previous stage.[35] That is, stages depend upon different qualitative states that discontinuously change as a whole. If each stage were continuous with the preceding stage—i.e., on a continuum—there would be one continuous whole with "stages" separated only by their quantity (on the continuum) rather than discrete wholes separated by their quality. Piaget clearly affirms the qualitative differences between stages and, hence, their discontinuity from one another.[36]

As noted in Chapter 2, however, many stage theorists and even some theorists trumpeting "discontinuities" in development do not explain these with discontinuous change. The reason is that change is thought to require time. That is, transitions between stages are assumed to happen along the line of time. Although change is undoubtedly accelerated during such transitions, some span of time—some quantifiable "length"—is still assumed. Because this period of time has to be crossed, it is therefore assumed that gradations "in between" stages are also crossed. Other discontinuities of change are described similarly. These are viewed as "spurts" of smooth and, hence, quantitative change rather than truly discontinuous and, thus, qualitative change.

The issue here is the same as the issue for causality. Particular forms of explanation have been outlawed in psychology because of their incompatibility with a largely unrecognized assumption—linear time. Quantitative change and efficient causation have been accorded "scientific status," and yet neither has been evaluated scientifically. In each case, *philosophical* decisions have been made with little or no awareness of the philosophical issues at play. Causation has been assumed to require linear sequence, and change has been assumed to require objective duration. Restriction to these assumptions does not, of course, negate the possibility of their validity for psychological processes. Nevertheless, their validity cannot be evaluated until they are identified and compared with alternatives.

Conclusion

A tempting, but inaccurate, conclusion is that there is a balance of linear and nonlinear approaches in psychology. After all, most Newtonian themes are met in this review with contrasting anomalies. For every explanation that assumes linearity or continuity, there appears to be an explanation that violates these assumptions. This conclusion might also seem supported by the number of "mixed" explanations in each subdiscipline: constructivist and linearist for developmental; rationalist and empiricist for cognitive; subjectivist and objectivist for personality; and of course, the extension of these mixes to psychotherapy. The problem with this conclusion is that the Newtonian elements overpower all such mixtures (with few exceptions). Regardless of the subdiscipline, the more "scientific" and "objective" elements dominate the mix.

This dominance is revealed in two ways. First, the sheer number of explanations issuing from the Newtonian framework far outweighs any competitor. Although no count of this sort has been formally undertaken, virtually all the theorizing of mainstream journals is founded upon assumptions of efficient causation and quantitative change. Indeed, most anomalous theories and theorists originate from fields outside psychology. Second, the robustness of the Newtonian framework, in the sense of its ability to subsume data, is staggering. This is a testament, of course, to its explanatory power. However, it is also a testament to its unrecognized status and its *perceived* explanatory power. Many theorists plainly do not see an option to the Newtonian framework.

This is probably best evidenced by the number of nonlinear theorists and constructs that have been "linearized" or cast as unexplainable. A scholarly discipline, such as psychology, does not construe theorists in this manner unless the assumptions underlying this construal are strong. Freud—a scholar of immense influence and countless disciples—has been repeatedly construed as a purely efficient causal theorist in many psychological texts. Jung— who specifically challenged objective time and explicitly labeled himself a teleologist—is routinely considered mystical, unscientific, and difficult to explain. These types of characterization extend not only to the theorists but the constructs themselves. Constructs that would seem to *require* a *non*-Newtonian framework (e.g., free will and self-transcendence) are either explained with antecedent determinants or excluded from the bounds of formal psychological explanation. Even family therapy, which has expressly attempted to escape its linear causal bonds, remains largely locked within the Newtonian framework.

The unparalleled strength of this framework is attributed, in part, to its unacknowledged status. If the Newtonian perspective were exposed more readily to view, a pluralistic discipline like psychology would immediately debate its advantages and disadvantages. As it is, however—grafted onto other pivotal concepts such as causation and change—this framework is rendered almost invisible. Heightening its visibility requires the identification of this grafting and the recognition of viable alternatives. This is the virtue of concepts like formal causation and qualitative change. They provide examples of causality and change that do not require a Newtonian approach to time. Effects happen and changes occur without the necessity of some absolute length of time.

These alternate assumptions also serve as the conceptual foundations for many anomalies reviewed. In most cases, the relevant theorists expressly deny linear-based assumptions and explicitly claim other forms of causation and change. Many of these anomalous theorists contend they are accounting for phenomena that cannot be understood adequately in a Newtonian framework—e.g., influences of the "future" and "unpredictable" events. From a linear perspective, of course, even considering the future to have a causal influence on the present is nonsense. Nevertheless, its lack of sense is based upon specific presuppositions of time. Alternative presuppositions exist that allow anomalies more viability and consideration. Providing these alternatives is the task of the final chapter.

Chapter 10

ALTERNATIVE ASSUMPTIONS OF TIME

The purpose of this chapter is twofold: (1) describe optional assumptions to those of the Newtonian temporal framework, and (2) provide some indication of their implications for psychology. As mentioned in the previous chapter, contrasting assumptions are badly needed. Any assumption that is as endemic to psychology as linear time cannot be meaningfully identified without other assumptions from which to distinguish it. If there is a Newtonian temporal "paradigm," then there must also be other paradigms worthy of consideration. Moreover, all the anomalies reviewed require conceptual grounding. Although some of the theorists behind these anomalies supply conceptual foundations for their explanations, these theorists have never been brought under one conceptual "roof."

This, then, is the first task of the chapter. Using the common themes of anomalies as our guide (from the previous chapter), two theoretical motifs are discernible. The first—*organismic holism*— takes as its rallying cry the holistic and nonreducible nature of time. Contrary to Newton, time is not separated from events of the world, nor are the dimensions of time—past, present, and future— separated from each other. Time is completely dependent upon the holistic relations that are found among events. The second theoretical motif—*hermeneutic temporality*—focuses upon the "temporality" of time. Unlike organismic holism (and absolute time), our *lived* experiences and practical activities are considered the foundation of temporality. This is not to say that time is subjective. From this perspective, conventional subject/object conceptions are

239

dissolved, leaving time as neither an empty container for events, nor an invention of the mind. Temporality is instead the articulation of the our world as we deal concernfully with its unity.

The second task of the chapter is to bring these unfamiliar assumptions to life in psychology. How can these alternative assumptions not only make sense out of the anomalies of the discipline but also account for the findings of mainstream researchers? This is obviously a tall order. In fact, this order is so tall that it cannot be filled at this stage of our exploration. The first step in this process, however, is to show how alternate assumptions of time *could* be applied to psychology. To this end, each of psychology's subdisciplines—as reviewed in the previous chapters—is discussed in relation to the two alternative frameworks. These discussions are brief, but they suggest the relevance and potential benefits of each alternative to the discipline as a whole.

Organismic Holism

Conceptual Roots

The conceptual roots of organismic holism—our first alternative to Newton's temporal framework—stem from four main sources: organismic biology, post-Newtonian physics, continental rationalism, and Hegelian metaphysics. These sources underlie many of the anomalies of the previous chapter. The organismic movement in biology, as led by Bertalanffy, provides the undergirding for anomalies associated with systemic methods (e.g., Saussure) and organismic family therapy (e.g., Minuchin). Post-Newtonian physics inspired several anomalous personality theorists, notably Kurt Lewin and Carl Jung. Rationalistic ideas have led to anomalies in developmental (e.g., Kagan) and cognitive explanations (e.g., Rychlak), with their attendant therapeutic applications (e.g., Rogers). And Hegelian influences have generally contributed to anomalous methods (e.g., Bertalanffy) as well as anomalous explanations (e.g., Piaget).

Given that these sources are the bases for many anomalies, let us examine them to see what type of temporal characteristics they may have in common. For example, it is clear at the outset that all four sources affirm a central principle. This principle of holism essentially asserts that *the qualities of entities are not in the entities themselves but in the relations among the entities.* Newton, of course, assumed that entities of the world exist independently in different regions of time (and space). Each entity has its own properties, separate from one another. Such entities can interact to form larger wholes, but they must cross the separate regions of time and space to do so. This means that the entities remain essentially unchanged.[1] Moreover, any study of their interaction ultimately reduces to the internal properties of the entities themselves.

Those who affirm the principle of holism deny this reductionism. They contend that synchronous relations among the entities are the crucial factors in determining the natures of the individual entities and the changes they undergo. Entities are not elements "first" and "then" wholes following interaction (as in a Newtonian framework). The wholes themselves—as entities related simultaneously—govern the qualities of the elements and, in some cases, their very existences. In this sense, an element's qualities are not linear or continuous with their objective past. They stem from the element's present relation to other elements, which itself can vary qualitatively from previous relations. Although the four sources of anomalies each endorses the principle of holism, each source emphasizes a different aspect of its character.

Organismic biologists, for example, emphasize nonmechanistic relationships. For them, the epitome of a mutual (and synchronous) interdependence is the living organism, one of the first metaphors for the principle of holism.[2] Although this holism was vital to ancient "biology," the rise of linear time has encouraged many modern biologists to adopt a mechanistic understanding of the organism. Cells are viewed as interacting across time and space, and thus existing independently of one another. Organismic biologists, on the other hand, have moved away from this type of Newtonian reductionism. Consider this passage from the organismic biologist Bertalanffy:

> While in the past, science has tried to explain observable phenomena by reducing them to an interplay of elementary units investigable independently of each other, conceptions appear in contemporary science that are concerned with what is somewhat vaguely termed "wholeness," i.e., problems of organization, phenomena not resolvable into local events, dynamic interactions manifest in the difference of behavior of parts when isolated or in a higher configuration, etc.: in short, "systems" of various orders not understandable by investigation of their respective parts in isolation.[3]

The rationalist Kant also emphasized the connectedness of parts. His notion that the mind imposes its own connectedness onto parts of the world was elevated to a new level in the work of the gestaltists. The principles of gestalt psychology are well known to most psychologists. Their dictum that the whole is greater than the sum of its parts is the principle of holism in different words. Their "laws" of organization (*pragnanz*) essentially delineate many of the types of holistic relationships that are possible. Indeed, these laws are thought to completely govern perception. The mind (or brain) is considered to organize all perceptual elements into wholes that accord with these laws. In this sense, the very existence of a "whole" is determined by its relation to the perceiver.

The holism of modern physics emphasizes a similar interconnectedness. The theoretical physicist David Bohm makes clear that many of his colleagues still subscribe to Newtonian reductionism: "The world is assumed to be constituted of a set of separately existent, indivisible and unchangeable 'elementary particles.' "[4] Nevertheless, he views both relativity theory and quantum mechanics as essentially holistic. Relativity theory teaches that "no coherent concept of an independently existent particle is possible," while quantum theory shows that electrons can be qualitatively different (wave versus particle), depending on their holistic context.[5] In both cases, physical entities cannot be understood except with the whole of which they are part. This means that entities which are normally considered to be separated in time (and space) are, in fact, simultaneously related. If these entities change, they change *as* a whole, and therefore change in quality, *discontinuously*.[6]

Discontinuous change, however, poses problems for our understanding. In fact, Hegelian philosophers argue that a new *dialectical* reasoning is required to understand holism because conventional deductive logic must be consistent (or continuous) in its logical flow. Conventional logic cannot contain an expression that does not appear in its premises. If Bob is a bachelor, then it does not follow that Bob is married. Bob must be consistent with who he is (in the premise), and a scientific law must be consistent with what it is across time (i.e., in prediction). If the earth is subject to gravity, then all apples will fall to earth. The conclusion that "all apples will fall" is implicit in the premise "earth is subject to gravity," and perfect Newtonian prediction is achieved. The problem is that discontinuous change is *in*consistent by this definition. An emergent property of the whole can occur (water) which is not logically deducible from the initial condition of its elements (hydrogen and oxygen). No linear prediction in this case is possible. A Hegelian type of nondeductive logic is therefore necessary to understand such nonlinear change.

These four "roots" of the psychological anomalies are all helpful in our quest for temporal alternatives. Each reveals an aspect of organismic holism that permits a better overall picture of this temporal assumption. First, time must permit simultaneous events to influence one another. Even "nonlocal" events—events supposedly separated in time and space—must be able to influence one another. Second, time must not separate the observer from the observed. As a part of the whole, observers must take themselves into account when explaining the observed. Third, time must be able to jump abruptly from one state (or place) to another. Because holistic change can be instantaneous, inconsistent, and unpredictable, time must also have these capabilities.

Comparative Description

Objectivity. Perhaps most importantly of all, time cannot be objective—at least not in the Newtonian sense. Newton's absolutism of time means that time exists apart from all the events occurring within it—"without regard to anything external."[7] Time is the container or medium for all events. It separates and distributes events across its spatialized "length," and endows them with its own special properties to some degree. Because time is continuous, for example, the events that occur in this medium must also be continuous.

Organismic holism, by contrast, cannot tolerate time's separation from events. From this standpoint, Newton's conception is considered highly abstract and speculative because time supposedly exists independently of the only things presented—the events themselves. Time in the organismic sense is the relation among events. Time is not an objective and separate medium, conferring upon the world certain absolutes regarding duration and sequence. Time is the connection of the world to itself, including the observer. As rationalists have noted, observers of the world have never been disinterested and objective—requirements of the Newtonian framework.[8] Observers are themselves related to the events being related.[9] In this sense, temporal characteristics are always relative to the context, culture, and observer.

Time, then, cannot be an objective measure with an equal (and uniform) set of units (an "interval" scale of measurement). Organismic time is better analogized to an "ordinal" scale in which only the *order* of things (in relation to other things) is important. "Order" in this sense is another term for relationship or organization. It is not objective (or absolute) and is always relative to the context, observer, and whole. It is, however, useful. Arbitrary times and dates can be formulated, implying before this and after this (relationally) as measured by some arbitrary device (e.g., a calendar or clock). Unfortunately, by organismic standards, the arbitrariness of this arrangement is often forgotten. Clock-time and the "order" of our world are not absolute and only make sense in reference to other events.

Linearity. Where then do we get this feeling of forwardness and the directionality of time? Newton, of course, answered this by postulating an unseen entity that is separate from events and endowed with "flow." If events are "in" time, from this perspective, then events have to be distributed in some linear manner.[10] This is the reason cause and effect have become so allied with linear time. Because a cause has a *logical* primacy, owing to its being the *producer* of an effect, Newton presumed that a cause must also have a *chronological* primacy. By linear definition, all primary events must occur in the past.

Organismic theorists, on the other hand, contend that primary events can occur in the present, simultaneously with the events they influence. Indeed, it is the overall (and simultaneous) gestalt of the "moment" that gives all events their qualities in the first place (see "principle of holism" above). Of course, a series of these momentary gestalts can themselves have a holistic structure. Momentary gestalts can occur together (in an order) so that they form a superordinate whole, with each "moment" as a subordinate whole (or part). This, then, is the "forward" movement of many natural processes from an organismic perspective—the invariant order (or structure) of momentary gestalts.

The directionality of human aging is an example of this. Each "moment" of the aging process is a gestalt of organismic processes, and yet other gestalts also follow this moment in an invariant order that is controlled by a static superordinate structure—our genetic code. From this perspective, aging is not a by-product of an independent entity called time, time is a by-product of aging. No objective standard (or objective past) is necessary. The events themselves imply a natural order and, hence, a temporal relation. In this sense, time simply means ordered change,[11] and no primacy of the past (linearity) is necessary. If anything, the ever-presentness of the natural structure that controls orderly change (e.g., genetic code) is considered to have the greatest influence.[12]

In another sense, holistic processes are not "forward" moving, even in this nonlinear and nonobjective manner. Momentary gestalts are always part of a larger whole that transcends conventional sequence and direction. Each momentary part—to be a "part"—must reflect the superordinate whole "within" itself. Because all parts derive their qualities from their relation to other parts (principle of holism), all momentary gestalts derive their qualities from their relation to other momentary gestalts (in the past and future). Even the smallest part of a hologram reflects the meaning of the whole. Analogously, a proper understanding of the qualities of any one "moment" is an understanding of all other moments as well. Each holistic moment has, in effect, the past and future implicit in its very identity. This is true even if the parts "across" time differ qualitatively. Qualitative shifts do not prevent their superordinate relatedness.

Continuity.   Unlike Newtonian time, organismic time is not consistent or uniform. Because time is not divorced from events, it is not the smooth and even measure of natural processes for which Newton had hoped. In fact, this is the reason that regular processes of change (e.g., clocks) were originally invented.[13] Natural events rarely occur in a regular (and mechanistic) manner, so regular processes of change had to be sought or invented to preserve the continuous notion of time. Indeed, from an organismic viewpoint, natural processes are momentary gestalts quite capable of emerging into completely new gestalts. Time must therefore reflect the holistic relationship among these gestalts, including the potential for *discontinuous change* (and linear inconsistency).

A variation of Hegel's dialectical logic may be of some aid here, because logic (or logos) is another way of talking about relationships. Conventional logic assumes uniformity of relationship and independence of entities in time.[14] To logically organize entities, one looks at the qualities of each entity (as though they inhered in the entity itself), and then puts them into logical categories based on the uniformity of their individual qualities (e.g., size and color). For example, cats and dogs are both categorized as mammals because of common individual properties (e.g., hair-covered bodies). Similarly, events in time are examined separately to understand their individual properties, and then related together as a process or empirical law when such events have sufficient uniformity (predictability, correlation). In either case—entities or events—conventional approaches to logic require us *first* to discern the inherent properties of the individual element before it can be placed into the correct category or law.

Hegel's dialectical logic, however, requires us *first* to discern the properties of the whole of which the element is part. This is because the properties of the part stem from its relation to the whole. A crucial property in this regard is the part's difference from other parts (as understood holistically). The property of "difference" necessitates mutual comparison (or contrasting relationship) because no entity can possess a quality of difference alone. Difference only occurs as an aspect of the "betweenness" of two or more things.[15] This is the heart of dialectical logic and organismic time. Unlike conventional logic and Newtonian time whose chief characteristics are uniformity and continuous flow, the chief characteristics of dialectical logic and organismic time are difference and discontinuity. Organismic change, in this sense, is relational difference (contrast, distinction, or opposition). An entity cannot change on its own; it must change *in contrast to* some other entity.

Organismic time, therefore, focuses less upon the uniformities and laws of reality and more upon the differences and contrasting relations among events—even if those differences and relations are discontinuous in nature. Discontinuity is only a problem if linear consistency and predictability are needed in explanation. Without this need, one can simply describe the relational (or contrasting) differences observed, e.g., Piaget's observation of discontinuous stages of development. A nonlinear type of predictability is possible because such observations may include repeated discontinuous relations (e.g., water from hydrogen and oxygen molecules). Time, however, does not flow continuously, any more than holistic changes flow continuously. Time hops and skips and flits, depending upon the differences among the natural processes of one's whole.

Universality. If time is discontinuous (or differential) by its very nature, then it follows that Newton's universality is not a tenet of organismic holism. Empirical laws (or universals) cannot retain their temporal relationships unless time itself is invariant. Discontinuities might mean potential qualitative changes in context which would disallow any absolute knowledge of time's future invariance. Newton, of course, argued that certain universals stand behind the changeableness of our world. Scientists, according to Newton, must train themselves to ignore the world as it appears and see the unchanging processes that supposedly underlie this superficiality. One of the ironic results of this is that scientific objectivity has come to mean ignoring the world as it is naturally presented and conjuring up unchanging processes that are never actually observed (e.g., force or gravity).

Organismic holism does not ask scientists to ignore the world as it is naturally presented. From this perspective, the relations among events must be taken seriously to discover their differential connectedness. Universalized theories and hypotheses cannot be useful. They are based upon deductive consistencies (and similarities) that would rule out potential emergent relations and qualitative changes which occur naturally. On the other hand, this has not stopped many structuralists and rationalists in this tradition from postulating "deep" structures that supposedly do not change with time (e.g., Chomsky and Bertalanffy). This apparent contradiction is resolved to some extent by recognizing the temporal assumption of organismic theorists. Universality across organismic time does not presume consistency and lawfulness in the linear (and deductive) sense. Organismic universality presumes differences and emergent phenomena in the dialectical sense. In other words, the universal "deep" structures that organismic theorists postulate (and use in "prediction") include and subsume data that would technically be unpredictable in a linear framework.

Reductivity. Reductionism is obviously incompatible with organismic holism. Because the qualities of an organic element stem from its synchronous relation to other elements (the principle of holism), no reductionism is permitted. This also disallows linear approaches to holism (e.g., interactionism). Linear assumptions require the repeated interaction of elements (e.g., mind and environment) to be sequential rather than synchronous in their relations. Information from one element must *first* travel to the other (across time and space) to affect it *later* in time, and *then* (still later in time) this "effect" must travel back, and so on (e.g., Bandura's reciprocal determinism and cybernetics). This means that a mechanistic or linear whole—the interactional process—is literally never a whole at any particular point in time. It is forever a reduction of itself.

Organismic holism, on the other hand, assumes that the whole cannot be reduced in this manner. All "elements" are related and reflective of the whole of which they are part.[16] Wholes occur at two levels: (1) the gestalt of all elements at the "moment," and (2) the superordinate structure (order) of momentary gestalts. Because the superordinate whole can occur across several clock-moments, it might seem as if this whole is subject to the same reductionism as linear approaches. That is, only a reduced portion of the whole would seem to be available at any point in time. This presumes, however, that time is linear in nature, with only "points" of time and durationless "instants" available as windows to the superordinate whole. Recall that a momentary gestalt of organismic time is not so limited (see the "Linearity" section above). Organismic moments incorporate the past, present, and future, and thus reflect the superordinate whole within themselves. This quality of organismic time precludes linear reduction.

This is not to say that organismic holists *never* reduce. In fact, a reduction of sorts is commonplace among these thinkers. Some would even maintain that a focus upon a part is important or unavoidable. The difference between linear and organismic reductionism turns again on the principle of holism. If the part is seen as *containing* its own qualities—whether through interaction or its own "internal" characteristics—this viewpoint is a linear reductionism. If, on the other hand, the part is seen as deriving its qualities from its fellow parts (in simultaneity), this viewpoint is an organismic reductionism. The part can be reductively studied, but it can only be studied in light of the whole from which it obtains its "partness."

Conclusion

Organismic time, then, is the relationship or connection *among* events. Time is not outside the events as an independent and self-sufficient entity. It occurs *through* the events as their relationships and changes occur. This relationship is not exclusively one of similarity (or continuity), as in more traditional explanations, but relies instead upon dialectical contrast (or discontinuity). Any directionality of these contrasting relationships stems from the organization implicit in the whole—what Bohm calls the unfolding of "implicate order."[17] Directionality in this sense is never absolute but always relative to the holistic nature of the events themselves.

Time is also nonlinear in the sense of nonadditive. A line metaphor for time gives all events a summative quality. To know a particular location on a line (point in time), one has only to add together the various line lengths (previous events) leading up to that point. In this sense, the present is always accounted for by the aggregate of past events (as stored or recorded). Organismic time, however, cannot be summated or aggregated because it is not viewed as an absolute interval scale (like a line). As noted above, organismic time is more analogous to an ordinal scale. The relation or order among events is more important than any individual additive qualities. The present can never be explained by summating the events of its past, because each event does not have its own properties divorced from other events of the past, present, and future (to be summed). Each "moment" of time is instead a gestalt of events that can emerge into a new whole inconsistent with its linear past.

Criticism

One of the most common criticisms of organismic holism concerns the placing of boundaries around wholes.[18] The extreme position— or the ultimate holist position—is that *everything* is part of a greater whole. If this is so, any treatment of an entity without reference to its totality hinders our understanding of its partness. Knowledge derived from a subordinate whole is potentially defective because we may not properly understand its role within the superordinate whole, and so on. This means that a clear understanding of the totality is crucial to any understanding of a part (or a part of the parts). The problem is: How does one define the totality? How does one obtain a grasp of the holistic quality of "everything"?

The theoretical physicist David Bohm appears to recognize this problem. Indeed, his own holistic approach to physics seems to pose this problem, yet Bohm is undeterred. First, the problem does not vitiate the need for understanding the holistic relation among parts. Parts are still parts of wholes and should be understood as such, regardless of whether we have a final answer to the identity of the totality. Second, Bohm seems to doubt whether we need a final answer to proceed as holists. Doesn't the nature of the "all" require a constant reworking and reformulating of our understanding anyway? He puts it this way:

> In our proposed views concerning the general nature of "the totality of all that is" we regard even this ground as a mere stage, in the sense that there could in principle be an infinity of further development beyond it. At any particular moment in this development each such set of views that may arise will constitute at most a *proposal*. . . . This proposal becomes itself an *active factor* in the totality of existence which includes ourselves as well as the objects of our thoughts and experimental investigations. . . . Through the force of an even deeper,

more inward necessity in this totality, some new state of affairs may emerge in which both the world as we know it and our ideas about it may undergo an unending process of yet further change.[19]

Thus, Bohm seems to view the lack of a *universal* totality as a fact of holistic life. He describes a holistic approach that he feels does not require an all-encompassing, once-and-for-all answer. Indeed, the involvement of ourselves in any postulation of the totality makes such postulation "unending." However, none of this detracts from the holistic nature of events, or the holistic nature of time as the ordered relation among such events.

Application to Psychology

What does all of this have to do with psychology? It seems important at this juncture to translate this alternative view of time into the subject matter of psychology. We should not, of course, overlook relevant anomalies to the Newtonian paradigm. Holistic anomalies exist in virtually every subdiscipline of psychology, and thus provide "living" illustrations of how psychologists employ organismic time. Nonetheless, a more explicit application to psychology would seem to be in order. This would help flesh out the "comparative description" above as well as facilitate critical review and future research on the advantages and disadvantages of each temporal alternative. An attempt is therefore made to delineate broad, though brief, implications for each of the subdisciplines reviewed in previous chapters.

Development. With organismic holism, developmentalists have no objective scale with which to mark or calibrate human development. Developmental stages may occur in a certain order, but organismic theorists do not view this order as part of a linear flow. Developmental order is certainly not linear in the sense of one stage *causing* (or even being deductively consistent with) the stage following it. Each stage is a discontinuous gestalt, and the order of their appearance (as they occur across dialectical time) is controlled by a higher (superordinate) gestalt (e.g., genetic structure). Universals are applicable as long as the context is similar (e.g., within a particular developmental stage). However, even the unchanging context of a particular developmental stage must be understood in reference to the superordinate structure of the life span of stages.

Otherwise, the stage's true meaning—its role in the whole—cannot be understood. Moreover, a move into the next stage implies a totally new context with a new set of rules. Knowledge gained during a prior stage of development may not be applicable. This is part of the reason predictability is so poor across the person's life span—people "emerge" discontinuously from prior phases of their lives with new meanings for old events and information.

Personality. Kurt Lewin's explanatory framework appears to be the prototype of a personality theory based upon organismic holism. Personality is not based upon an accumulation of past experiences, nor is personality isolated from the "field" of interpersonal factors in which it is embedded. Instead, personality involves factors of the present interacting synchronously with one another. No influences upon personality exist in separate regions of time and space. All are viewed as parts of a greater whole, important with respect to their position in this whole rather than any internal qualities they might seem to possess. In this sense, the traditional focus upon the individual in personality theory is reductionistic. Traditional approaches treat individuals as quasi-independent entities, caused more by their internal characteristics (e.g., traits, intrapsychic dynamics, or learning histories) than by their relation to other individuals. This is the reason that holists, such as Lewin, have favored understanding individuals in groups. Even groups without a mutual history (e.g., group therapy) still form an invaluable present context for understanding normal and abnormal functioning.

Method. The usual view of method is that it isolates particular variables (independent variables) to understand their influence on the processes under study. This isolation or "independence" is not considered possible from an organismic perspective (see Chapter 4). A "variable" cannot be separated from its natural context because crucial aspects of its identity could be lost. Even the standard sequencing of independent and dependent variables is suspect. This sequencing implies that each variable occurs in a separate region of time and space, preventing study of synchronous causal relationships (e.g., figure-ground). Methods based on continuity of time are also questioned. Experimental designs that only sample a process on two occasions (before and after) could miss unique discontinuities and qualitative changes. Such changes also jeopardize conventional notions of replication. Natural processes could emerge into a new form and not be replicated under the old set of experimental conditions. From an organismic perspective, this does not mean that such processes are less real or empirical. They simply are not universal in the linear sense.

Cognition. Past events or experiences drive nearly all mainstream models of cognition. This includes the recent past (immediate stimuli), the distant past (stored input), and any interaction of the two. Organismic holism clearly challenges these explanations of cognition. If stages of "information processing" are involved, they are likely to be discontinuous (rather than continuous) in nature. If "past" experiences are involved, they are part of a present gestalt of meanings that can totally escape the past as it was originally input. That is, old meanings can be qualitatively different, depending on their present arrangement. Of course, the notion of "input" is itself based upon a sequencing of information. Holists argue that the mind and its environment are themselves parts of a simultaneous structure. No information "flow" is necessary because the two parts "communicate" with each other by virtue of their relation as parts of a superordinate gestalt. Cognitive experimentation—typically conducted in the laboratory—also lacks the natural context for truly understanding this gestalt.

Individual Therapy. Most individual therapies are reductionistic from the standpoint of an organismic theorist. Especially if therapists concentrate upon qualities "inside" the individual (stored input, intrapsychic dynamics, physicochemical processes), they are ultimately reductive in the Newtonian (and mechanistic) sense. Still, a holistic form of individual therapy can be practiced. This treatment must "begin" with the whole of which the individual is part—family, social system, and culture—and "then" understand the individual in this light. Otherwise, vital qualities of the individual are overlooked. In addition, the contextual whole of patients (and their individual qualities) are subject to dramatic shifts because wholes can shift discontinuously. This has three important implications. First, the therapist's knowledge of the whole must be constantly updated. An instantaneous shift in holistic qualities can occur without warning because it lacks any past events "leading up" to it. Second, this shift can occur without the therapist's knowledge. Most therapy takes place in the therapist's office, preventing any direct observation of the client's "natural context" that is integral to the shift. Third, traditional assessment of the individual is inadequate. Many diagnostic categories and assessment procedures assume that individuals do not change *qualitatively* across time.

An organismic individual therapist, however, must remain open to the inevitable discontinuity of the patient's "condition." This change may occur only for a "moment." But such change may be important to a proper dialectical understanding of how that moment differs from other moments.

Group Therapy. Linear conceptualizations of group therapy maintain psychology's traditional individualism and reductionism. Again, as long as patients are seen as containing or causing their own qualities, the focus of therapy eventually turns to those self-contained factors (see Chapter 7). Organismic group therapists, however, focus upon the relational qualities of the group—the "groupness" of the group. This groupness is not a linear holism in which individuals are viewed as interacting through forces (e.g., sights or sounds) that travel between the separate time and space regions occupied by the members of the group. This groupness is an organism in the classical sense of a single, simultaneous (here-and-now) unit of life. This does not necessarily mean a "groupmind" in the same sense as a human mind, but it does mean an emergence of group qualities that is separate from the summation of its individuals. Emergent qualities also extend beyond the group. Because the group is itself part of a larger group (society), effective group therapy must take into account this larger group. This also means that any change in the smaller group can reflect holistically upon the larger, with a potential beneficial (and simultaneous) effect upon society as a whole. Universalized theorizing is not ruled out here. Such theorizing must embrace rather than avoid discontinuities, but underlying group dynamics are often viewed as the "deep structure" of many groups.

Family Therapy. Family therapy is probably the closest subdiscipline to implementing organismic holism on a wide scale. After all, Bertalanffy—one of the primary sources of organismic holism—is also one of the primary precursors of family therapy theory. Nonetheless, as noted in Chapter 8, linear (or mechanistic) holism still dominates the field. Each family member is thought to be "outside" the other, with the therapist considered "outside" the family as a mechanistic unit. This is true no matter how "interactive" these elements are (e.g., second-order cybernetics). All interactions involve linear sequences, and so each part of the system "affects" the other from a separate region of time and space. This also means that family therapists must see several phases of this sequence before they can identify the overall system. Organismic therapists, on the other hand, view families as having at least two levels of system: the momentary system (all members in simultaneity) and the superordinate system (momentary gestalts across dialectical time). *Both* levels are thought to be reflected in any *one* "moment." Admittedly, a proper understanding of the superordinate system is difficult from one momentary gestalt, but both systems are present, and holistic intervention into either is possible. Therapists, in this sense, do not intervene from the "outside" but attempt to alter their simultaneous part of the whole in a manner that effects discontinuous change.[20]

Hermeneutic Temporality

Conceptual Roots

The conceptual roots of hermeneutic temporality stem primarily from philosophy. Whereas the sciences (such as biology) mainly inspired organismic holism, the humanities (such as philosophy) mainly inspired hermeneutic temporality. Interestingly, the diverse origins of these two alternatives have not prevented considerable conceptual overlap. Their mutual rejection of the temporal framework represented by Newton has led both to champion holism and qualitative change. However, they both do so for different reasons, and their alternative conceptions are fundamentally distinct.

Hermeneutic temporality essentially stems from three primary sources: phenomenology, existentialism, and hermeneutics. As with organismic holism, these sources form the conceptual basis for many anomalies reviewed in the earlier chapters. Indeed, these sources underlie most *other* anomalies—those that are not founded upon organismic assumptions. Phenomenology provides the rationale for many personality (e.g., Jung) and therapeutic (e.g., Kohut) anomalies to mainstream conceptions. Existentialism is the theoretical foundation for anomalies in individual therapy (e.g., May) and group therapy (e.g., Yalom). And hermeneutic conceptions have led to recent anomalies in method (e.g., Gadamer), cognition (e.g., Dreyfus), and family therapy (e.g., Whitaker).

One of the fundamental concerns of all three conceptual roots is the issue of time.[21] Although there is no unanimity on all aspects of the issue, there is general agreement that a Newtonian approach is counterproductive. Perhaps the most incisive critique of this approach, as well as the most extensive reformulation of time, occurs in the work of Martin Heidegger. He proposed a hermeneutic form of time. As it happens, Heidegger was closely associated with all three conceptual roots. He is considered one of the founders of hermeneutic philosophy as well as a major contributor to existential philosophy.[22] He is also thought to have further developed (and explicated) many of the ideas of his mentor, Edmund Husserl,[23] the so-called father of phenomenology. Consequently, we will briefly outline Heidegger's view of time—or "temporality," as he called it— and use his conceptions as the basis for our presentation of this second alternative to Newton's temporal framework.

Traditional Metaphysics. Heidegger intended to begin his philosophical study at the very beginning—the study of *being*. He knew that "being" is one of the oldest and most essential issues of Western philosophy.[24] Unfortunately, from his perspective, most philosophers accept the traditional modernist conception of "being" without examination. Beings—humans, gods, objects, and so on— are considered to imply thing-like entities that exist independently of human consciousness.[25] Although gods may be immaterial beings and humans may be material beings, all beings are objective entities in the sense that they do not depend upon our consciousness for their existence. Heidegger noted that virtually all explanations are based upon this conception. Indeed, all external reality is now construed as a set of objective entities, independent of our consciousness. This is the metaphysic in which time itself—supposedly an aspect of that "external" reality—is considered an objective entity outside our consciousness.

Rene Descartes perhaps best articulated the main implication of this conception of being: the world is split into two entities, the subjective and the objective. The inner world of the subject consists of our consciousness. Thoughts and feelings can only take place "inside" this world, the world occurring supposedly within the boundaries of our skin. The outer world, however, is the supposedly independent world of objective reality, a world of things (or beings) as they really exist. Of course, one of the first questions to arise with this split world is: How do the two worlds relate? How, for example, does the subjective world (where each of us resides) know about the objective world? We undoubtedly have transactions between the inner and outer worlds through the portals of our senses, but how well does our inner world *correspond* to the outer world? This became such a fundamental question of epistemology, that all the great epistemologists addressed it.

This "correspondence" notion of epistemology also became the foundation of Newtonian science. If our concept of a thing (in our inner world) is correct, then it must *correspond* to the thing (in the outer world). If prospectors believe (in their inner worlds) that they have found gold, their belief must correspond to the actual finding of gold (in the outer world) for this belief to be true. In this sense, scientific experimentation became the primary means of testing for the truth of ideas. The ideas themselves—theories and hypotheses—are part of the inner, subjective world of the scientist. Hence, these ideas are considered mere opinion or speculation. They must be tested against the outer world of objectivity—the data of beings—to see how well they correspond to reality as it really is. In this sense, the need for correspondence (in both science and epistemology) results from a conception of being that automatically presumes the separation of subjects and objects.

Hermeneutic "Metaphysics". Hermeneutic metaphysics, as inspired by Heidegger, does not accept this separation. In fact, Heidegger disliked the term "metaphysics" (preferring ontology), partly because this branch of philosophy has been so long associated with this separation. Heidegger felt that the conception of being which brought about these implications in science and epistemology was a false one. Being is neither an objective thing nor a subjective idea from his perspective. To even begin to conceptualize a human "being," he reasoned, the subject/object dichotomy must be transcended altogether. The problem is that this dichotomy is so endemic to Western thinking that it is difficult to excise from our explanations. Indeed, to remove this dichotomy totally, we must recognize the "mode of engagement" that spawns it.[26]

According to Heidegger, the primary mistake of modernists, such as Descartes, is that they rely upon a "present-at-hand" mode of engaging in the world. This mode requires a *detachment* of ourselves from ongoing practical involvement in the world. In this mode, we "step back," reflect, and turn to general and abstract (i.e., context-independent) conceptual tools, such as logic and calculation, to solve problems. In short, we *theorize*. Modernists view this type of theorizing as the pinnacle of intellect. Unfortunately, what modernists may not realize, Heidegger observed, is the influence of this mode of engagement on their philosophy. To consider context-independent ideas as the height of human intellect is to have a philosophy that reflects this independence. Therefore, being—as considered theoretically (and universally)—is separated from the context of its existence, including the being interpreting the being.

Heidegger argued, by contrast, that theorizing is only one of several modes of involvement with our surroundings. Indeed, theorizing is not even the best or most significant source of knowledge, because it is itself derived from more basic modes.[27] Primordial and direct access to human phenomena stems from a "ready-to-hand" mode. As Heidegger asserts, "the ready-to-hand is not grasped theoretically at all."[28] No abstract, context-independent awareness occurs while we are in this mode. Our "awareness" pertains only to context-filled activity, whole actions leading to particular goals—such as mailing a letter or hammering a nail. Many seemingly discrete movements and "objects" (letters, hammers) may be involved in such tasks, but we do not reflect upon them (nor do they exist) as independent theoretical entities in the manner of a subject/object dichotomy. They are, in a real sense, an inextricable part of us and the activity.

When this mode of engagement is viewed as fundamental, a whole new metaphysic or rather, ontology, becomes apparent. From this mode, being cannot be divorced from its surroundings or its interpreter/observer. Being is first and foremost involved in practical activities, before any theoretical understanding has taken place. Being cannot be understood apart from this involvement—apart from its particular surroundings, tools, and culture. All such factors must be understood as a pretheoretical and "situated" unity. They are fused into the practical activity in which all are engaged and intertwined. The subject/object dichotomy has no place here. Humans are not "beings in a bag of skin" with the world outside the skin, but beings-in-the-world with the world and being inseparably linked to one another—both parts of the same entity. Heidegger called this deeply contextual entity *Dasein*.

What does this new "metaphysic" imply for time? As the title to Heidegger's magnum opus, *Being and Time*, indicates, time holds a central role. Nonetheless, time cannot have a "being" independent of other entities (subjective or objective) in the traditional Newtonian sense. If ready-to-hand engagement is considered the primary route to understanding, then time must be part and parcel of *Dasein*. Time, in this sense, is not "filled up" with events. Time is the connectedness of meanings that is implicit in *Dasein*'s practical activities of everyday life. Heidegger prefers the term "temporality" for this connectedness because it better connotes the temporariness or "unsettledness" of *Dasein*.[29] Time in this temporal sense is not uniform and continuous but dynamic and variable. This sense of temporality endows knowledge with a "situatedness" that is implicit in ready-to-hand engagement.

Comparative Description

Objectivity. It seems clear from this analysis that temporality cannot have an objective "being." Newton's conceptions, of course, are rooted in the traditional metaphysic. He thought of an entity as having a "being" that is "outside" other beings, separate and independent. This inside/outside distinction is partly why Newton favored spatial metaphors—there is always some inside/outside (spatial) difference between entities. For the same reason, time is spatialized (in a line metaphor). Time preserves the traditional independence of beings by keeping the beings separate from one another *in* time (e.g., sequencing of family members). Of course, time itself is separate from beings, existing as Newton put it, "of itself, and from its own nature . . . without relation to anything external."[30]

Hermeneutic temporality, on the other hand, is integral to *Dasein.* In fact, it does not exist apart from *Dasein.* Heidegger held that entities are not separate from one another in objective time and space. They are all part of the emergent whole of practical activity, while temporality—the "movement" and "direction" of this activity—is not itself separated from the activity. For example, the "movement" of an involving (ready-to-hand) conversation is not independent of the conversation itself. The participants have a sense of "having been" somewhere and "going to" somewhere else, even if they do not know their exact destination. This sense of movement is neither separate nor objective. Temporality does not "span" the activity as a spatialized conception of time. Instead, temporality is inherent in every "moment" (or act) of the conversation.

Continuity. Hermeneutic temporality evidences a continuity of sorts, but it is definitely not Newtonian in nature. Because temporality does not exist apart from *Dasein,* it cannot have the Newtonian characteristic of pure continuity or "flow." Newton, of course, needed this uniform flow to provide an "objective" measure for phenomena. His aim was to actualize the detached form of knowledge originally envisioned by Descartes and other modernists. Their elevation of present-at-hand engagement meant that some type of context-free uniformity was available. This implied that time and space (two primary aspects of context) were themselves continuous and unvarying in their respective qualities.

Heidegger, however, discusses the difference between temporality and continuous time: "It is exactly when *Dasein* is immersed in everyday, concernful 'living along' that it does not understand itself as running along in a continuously enduring sequence of pure 'nows.' The time which *Dasein* allows itself has what might be called gaps in it."[31] This implies that temporality is open to the "gaps"—the discontinuities—of our concernful living. It is open to our ready-to-hand engagement without preconditions. If holistic change occurs—if a "leap" of insight happens or a qualitative change occurs without intervening states—then temporality can accommodate this. Indeed, because *Dasein* is itself a gestalt, our practical activities are *likely* to be experienced as discontinuities. As Andrew Fuller notes, "Lifeworld moments . . . discontinuously jump into definite form as this meaning or that."[32] Of course, we can move to a theoretical mode to *impose* continuity upon events retrospectively, but this is secondary to the more basic mode of engagement.

This is not to say that there is no sense in which ready-to-hand engagement is continuous. Like a good story, our practical activities are "drawn along" in a "direction" that provides a sense of connectedness.[33] Such connectedness is vital because the story would be random babblings without it. Still, the continuity of a story "line" does not have to rule out qualitative changes (or discontinuities by Newtonian standards). The context of the story could change completely, yet we would not have to lose our sense of "having been" and "going to" somewhere. Epic novels often exemplify this. With Tolstoy's *War and Peace*, for instance, we move from story to story (context to context) without losing the unity of the entire book. The storyteller, even in this discontinuous sense, is always headed toward something—a "future"—though we may not know this future until the story is over.

Linearity. The "futurity" of ready-to-hand activities brings us to the "linear" dimension of temporality. This is not to imply that temporality has this property in common with Newton's linearity. Newton viewed time as an immense line stemming from the past and heading into the future. All moments (or points) of time are durationless and empty. Past, present, and future are separate dimensions as they occur in sequence along this line but are nevertheless parts of the same flow and thus remain consistent with one another. As noted in Chapter 9, this means that only quantitative (and consistent) change is possible from one moment to the next.

Hermeneutic temporality has none of these linear characteristics. The primary reason is that it does not attempt to cut across situated moments of practical activity. No spatialized "length" or "distance" or "quantity" of temporality is recognized as fundamental.[34] Instead, the present "moment" of the activity is the focus. The present is pregnant (rather than empty) with the meaning of the activity, partly because the present dimension of time is not separated from other temporal dimensions. In Newton's framework, the present is the "knife-edge" separating the past and future. In Heidegger's framework, however, no such separation occurs. The "present" *implies* the past and future in the same meaningful moment. In fact, according to Heidegger, the present cannot be meaningful without the presence (in the present) of the past and future. Just as "being" cannot be understood apart from the context of the world, the present cannot be understood apart from the context of the past and future. A conversation among friends, for example, cannot be understood in terms of the linear present only. The meaning of the friends' past relationship as well as the import of their perceived future together are crucial to the "nowness" of the conversation.[35]

This means that a clear directionality or lived "arrow" of temporality is involved in all ready-to-hand activity. The integrated present is experienced as stemming from the "past" and headed toward the "future." Indeed, it is the reification of this experience that led historically to the Newtonian view. This directionality of present activity, however, does not necessarily imply a "flow" across instants of time nor does it imply the type of continuity (and consistency) postulated by Newton. These properties originate from present-at-hand theorizing about practical experience. The context-free quality of this mode of thinking leads to the notion that the presentness of the activity is itself free of past and future context (since they are all supposedly separated through linear sequence).

In hermeneutic temporality, by contrast, each moment is filled with temporal context. In this sense, *Dasein* can change quite radically as the context itself changes. This is true even of temporal dimensions. Newton's conception leaves the past "dead" and immutable, while it constrains the present and future because of their consistent "flow" with one another. Temporality, on the other hand, permits all dimensions of time to affect one another (e.g., the "future" affecting the "past"). The past is not dead and gone but part of the living present and future. Our memories are not "stored" and "objective" entities but living parts of ourselves in the present. This is the reason our present moods and future goals so affect our memories (see Chapter 5). Unlike the consistency of Newton's world where everything is mechanistically determined (and part of the same determining "flow"), Heidegger postulates a never-ending *possibility* and mutability that is inherent in *Dasein*.

Universality. This type of ever-present possibility is, of course, incompatible with Newtonian universality. Given the modernist notion of being, Newton affirmed the importance of scientific objectivity and the triviality of "surrounding" context (as extraneous variables). This led him to ignore differences among contexts and concentrate instead upon similarities among the "objects" under study, as if they were enveloped in a contextual vacuum (and excluded physical, temporal, and interpretive context). This led him to postulate "laws" that supposedly stand behind and govern these objects. The flux of the contexts and objects (as the scientist experiences them) is viewed as superficial—i.e., "appearance" only. The scientist must look behind and across the events of this flux to see the universal principles underlying them.

Hermeneutic temporality does not look "behind" or "across" the integrated "moments" of ready-to-hand activity. The practical activity of everyday life is taken for what it is, *as* it is experienced. Only the supposedly erroneous ascendance of present-at-hand reflection would lead to the valuing of knowledge that has nothing to do with what is actually experienced. This explains why positivists—who champion observables as knowledge—postulate universal "laws" which themselves are not being observed (e.g., force). Such universals require relations among events—relations across the line of time—that are not themselves experienced at any one point in time. These universals also overlook the experien*cer*. After all, it is the person's selective attention to an event, making it a figure against the ground, that gives it "event" status in the first place. To then study a series of events as though the event-grantor were not present is to make the whole process of study unintelligible by hermeneutic standards.

How then is the process of study made intelligible? For the hermeneuticist, this process always begins with what the action is—as it is experienced. Note that the object of study is not a "thing" but an integrative action. That is, the ready-to-hand *pre*reflective understanding of the action is taken as the ground of knowledge. This ground does not include any detached reflection. It does not include consideration of the action in a contextual vacuum or a transcendent relationship with other similar actions across time. Instead, the action is understood with its physical, temporal, and interpretive context in the "now." A question, however, often arises concerning this "now:" If all fundamental knowledge is centered in the present, how can the unity of knowledge be achieved? With all knowledge related only to the now, no knowledge would seem to be applicable to other moments (i.e., universal).

To answer this question, we must first note the linear notion of time it presumes. Linear theorists view the three dimensions of

time as occurring in sequence, each dimension forever separated
from the other two. Although this is widely accepted, it prohibits
the past (or future) from being a simultaneous context for the
present. Knowledge limited to the present moment can never be
applicable to any other moment. Indeed, this is the main impetus
for scientific "laws." Such laws are the primary means of cutting
across or transcending the various moments of time. Causal laws,
particularly, are intended to bring the context of the past into the
present through determinism. The problem is that determinism
only brings the immediately preceding event (of the past) into the
present. The past as a whole—certainly the past as a gestalt—
never accompanies the present, by definition.

Hermeneutic temporality, however, views time differently. Any
meaningful understanding of the present is necessarily an understanding of the past and future (simultaneously). No transcendence
of moments across linear time is required. A type of relevant (and
universal) knowledge is formulated *in the present*. This is not to
say that a meaningful whole cannot occur across several moments
of clock-time. The "living now" of hermeneutic temporality is not
constrained to the durationless instant of linear time. Several parts
of a practical activity can be united in meaning. A spoken sentence,
for example, can possess a unified meaning, despite its sequence of
utterances across moments of clock-time. No Newtonian universal
is necessary to explain this. The meaning is inherent in the activity
rather than in some law of meaning that lies outside or behind the
activity.

Reductivity. Temporality is entirely holistic, though some aspects of its holism may be subtle. Newton's reductionism results
from two supposed facts: (1) All events of a process are distributed
across time, and (2) direct access to the process occurs only one
moment at a time. Thus, Newton could never directly observe all
the pieces of the process at the same time. He could, of course,
observe each piece separately, record it in some manner, and then
attempt to view all recordings together (in his memory if no other
place). Nonetheless, this piecemeal process would always keep him
from knowing how (or whether) the pieces related as a whole. A
holistic (and simultaneous) relationship might change their function or even identity, yet a piece-by-piece recording could never
capture this.

The hermeneuticist's focus upon the present may seem similarly reductive. The contextual nature of this present would not

appear to save hermeneutic temporality from the same problem as Newtonian reductionism. Processes across time would seem to be understood in the same piecemeal fashion, ruling out access to all parts simultaneously. Once again, however, this problem of reductionism is based upon a linear notion of time. Because the hermeneutic present is not distinct from the other two dimensions of time, its "reduction" to the "now" incorporates the past and future. Only a linear formulation omits these from the present, since they are separated in time. A hermeneuticist views the present as a holistic microcosm of all time (and space). No reduction of any sort has occurred. This problem of reductionism also assumes that universal processes—processes that occur across time—are the processes of interest (truth). Hermeneutic temporality considers truth to be an integrated "nowness." That is, truth is what "is," not what is theorized to be "behind" or "across" events.

Conclusion

Hermeneutic temporality violates all five of Newton's characteristics of absolute time. Nevertheless, temporality also offers a reformulation of each characteristic that explains its conceptual origin in Western society. First, temporality has an objectivity of sorts. It does not assert any independence of consciousness, but it does contend that knowledge can be shared and communicated (through culture and language). Second, the connectedness of temporality evidences a type of continuity. *Dasein* is in the midst of a connected narrative—coming "from whence" and going "to whence"—though this connectedness is quite at home with what a Newtonian would consider a discontinuous change in that narrative. Third, a type of linearity of time is revealed. Temporality offers direction and gives importance to all three dimensions of time, yet this directionality and dimensionality pertain to the meanings inherent in practical activities rather than an independent medium for events.

Fourth and fifth, some form of universality and reductivity is important when ready-to-hand engagement is presumed to be the highest form of knowledge. To be sure, there is nothing free of context—nothing universal in the Newtonian sense. There also can be no reduction to the linear present because this temporal dimension has no meaning apart from the context of the past and future. However, a type of hermeneutic universality is possible because the meaningful "now" is itself a microcosm of time. To know this present fully is to know a "universal" in the sense of being applicable to all time dimensions in the "moment." This universal, though, remains "situated," and thus all dimensions are open to reformulation and "possibility."

Criticism

Two criticisms of hermeneutic temporality are common: relativism and subjectivism. The first holds that a temporally situated interpretation ultimately reduces to an absolute relativity, and absolute relativity is a denial that any truth is possible. This is a common misconception of Heidegger's position.[36] Certainly, Heidegger is arguing that there are no universal truths. Nonetheless, his denial of the universality of truth (and time's linearity) is *not* a denial of truth per se. If anything, Heidegger is attempting to understand the truth as it truly presents itself to us—in a situated context of the "living now." Moreover, he considers any "truth" that transcends the present context (e.g., universal theorizing) to be difficult, if not impossible, to apply to the unique context in which we all live.

The second criticism is a variation upon the first. It asks how we can hope to test or discern the truth in this situated present. How can we know if something is true or false within a situated present? Is this not total subjectivism? First, this is a Newtonian question because temporality can only be subjective if we presume subject/object distinctions. Second, Heidegger's denial of an absolute standpoint for such a test does not deny the possibility of legitimate ways of discerning truth or falsehood. His denial only challenges conventional views of truth (universal, certain). Packer illustrates this with a man offering a woman a flower.[37] Although many interpretations of this action are possible (e.g., peace offering or a gesture of appreciation), the context is such that it is unlikely to mean a threat or a dental appointment. In short, we can make decisions about truth in the usual manner. We just have to give up the notion of objective certainty and universal applicability (see Chapter 4).

Application to Psychology

Let us now turn to the implications of hermeneutic temporality for psychology. The existing hermeneutic literature seems to devote the most space to the more basic aspects of psychology, such as method and mind. These are the topics that first held the attention of the philosophers who articulated temporality. Consequently, these are the first topics to be developed in psychology. There are, however, important implications for every subdiscipline that we have reviewed.[38] A broad sense of these implications is described below.

Development.   The "past," in the sense of development, is highly valued in hermeneutic temporality. Development is considered to be a vital portion of the necessary (and preconceptualized) background to any practical activity. Indeed, the "historicity" of the situated "now" is discussed more in the hermeneutic literature than its "futurity,"[39] though the two temporal dimensions—past and future—have equal contextual import for the developmental enterprise.[40] There is a movement and a direction to life itself, according to Heidegger. Each human in their "being" has a life story, a feeling of connectedness, and a "developmental" account of this connectedness is a narrative in natural language form. An overall pattern may emerge from such an account, but the pattern is laden, rather than free, of context. This would allow both continuities and discontinuities in the Newtonian sense. No linear predictability would be necessary in such a narrative, but some predictability might be expected since even meaningful wholes may change only slightly (or not at all). Developmental "progress" might be less emphasized. Most notions of progress stem from linear manifestations of accumulation and temporal superiority (later is better). Moreover, an adult mode of thinking is typically considered to be abstract, logical, and present-at-hand. The ability to think in this manner would not necessarily be viewed as progress toward a higher mode of cognitive development. As noted above, this mode of engagement is not thought to be the most valid means of attaining knowledge. It is one among many means, with perhaps a child's more holistic and concrete mode being the more fundamental.

Personality.   Carl Jung may be the closest of traditional personality theorists to the hermeneutic perspective. He explicitly identified his own methods as hermeneutical (see Chapter 3) and emphasized the synchronicity of past, present, and future in many of his conceptions. In fact, his notion of archetypes is an excellent example of cultural influence and tradition in the "now," without that culture and tradition *determining* the "now" (as in a linear frame). There is always a sense of the "possible" (and temporal) in Jung's writings. Unlike the vast majority of personality theorists, he emphasized the goal-directedness of the future (in his teleology) as encompassed in the "now" of the living present. Many of his followers reify his theoretical constructs, but Jung himself sought to avoid this. His theoretical constructs are not "structures" in the Freudian sense, but *meanings*. He was the first to advocate their abandonment if they prevent the therapist from truly encountering patients where they "live." However, despite Jung's efforts, the concept of "personality" has come to imply that a person houses a set of attributes or traits which remain fairly continuous across time. Shweder and Bourne, for example, have shown that Western thinkers prefer a context-free understanding of others.[41] They tend to describe a person as possessing traits, such as honesty or esteem, which remain stable across diverse situations.[42] Like Jung, though, a hermeneuticist sees this as reifying interpretations of the person—accepting the traditional metaphysics of objective entities. From the perspective of temporality, any personality attribute has as much to do with the interpreter as the interpreted.

Method.   Method can no longer depend upon the correspondence theory of truth (see Chapter 4). No longer can truth be discerned through the correspondence of external facts and their mental representations. First, there are no "external" entities because humans are considered to be "outside" with the entities themselves.[43] Second, no "fact" is free of presupposition; indeed, no method is free of presupposition. One of Heidegger's students, Hans-Georg Gadamer, outlined what he called "the prejudice against prejudice."[44] This is the scientific conviction that truth requires the eventual removal of all presuppositions so that objective facts can be obtained. To a hermeneuticist, however, "prejudice" (preconceptual understanding) is always present and always must be present to make anything intelligible. The meanings of the living present must be taken as they present themselves. The brownness of the chair belongs to the chair itself. It is not viewed as a "subjective" sensation provoked by "objective" processes impinging upon our sense organs. Indeed, the chair is not limited to its various sense qualities. It is—from the first—something to sit on. In ready-to-hand activity, "things" are seen more for their *meaning* (purpose, significance) relative to an activity than for their "objective" perceptual qualities. This obviously de-emphasizes quantification, lawfulness, and controlled experimentation and emphasizes qualitative description, analogical understanding, and narrative modes of description.

Cognition.   In mainstream cognitive models, the subject/object distinction reigns supreme. In fact, from a hermeneutic perspective, current formulations of cognitive science might not *exist* because they assume that cognition can be an "object" of study. This is not to say that modern cognitivists do not emphasize the mind's interaction with the environment. However, even the term "environment" (as borrowed from the natural sciences) connotes an object-like reality which can reach cognition only by traveling across time and space. Hermeneutic approaches, on the other hand, center on lived meanings that are not accessible to present-at-hand reflection. The past, in this sense, is never objectively stored in the mind. The past is integral to the present as part of the contextual interpretation of "situated" circumstances, including the present image of the future. Erving Goffman, for example, speaks of a person's past as an "apologia." This apologia creates an "image of a life course . . . which selects, abstracts, and distorts in such a way as to provide him with a view of himself that he can usefully expound in *current* situations."[45] The construction of an apologia is not unlike an author's construction of a biography for a new fictional character. The biography is not objective or stored for retrieval; it is constructed in light of the ongoing story. This construction never ends (though it is halted in death), and as Dreyfus notes, portions of it—the preconceptual portions—are never accessible to the cognitive researcher.[46]

Individual Therapy.   What does this mean for individual therapy? Perhaps foremost, it means that an individual therapist does not treat what some call the "self-contained" self.[47] That is, "containers" of dysfunction, such as cognition, personality, and behavioral repertoire, can no longer be the center of therapy.[48] *Dasein* must become the center of therapy. As Jerome Frank says, patients bear a striking resemblance to a literary text—short on objective facts and liable to many interpretations.[49] In this sense, therapists should understand their own "presuppositions" in interpreting this text (as much as possible). This would facilitate their "openness" to the patient "as they really are" and help avoid the dual temptations of reifying the client's history and endowing the patient with attributes or diagnoses. At some point, there must be a "communion" with *Dasein* to understand the patient truly. This is not a communion of inner worlds, crossing separate regions of time and space to "reach" one another. Hermeneuticists assume instead that considerable preconceptualized common ground already exists in culture and language to enable understanding. What does the therapist do with this understanding? How is *Dasein* best "directed" as a future possibility? The closest Heidegger comes to this topic is his discussion of authentic and inauthentic modes of being-in-the-world (specifically being-with-others). Authentic (ready-to-hand) engagement in the world entails concerned and practical involvement with others. Inauthentic being, on the other hand, treats one's self and others as "objects" forever estranged from the "subjectivity" of one another.

Group Therapy. Group therapist, Irvin Yalom, asks himself how this authentic human encounter can be facilitated.[50] Much of his answer lies in his formulation of group psychotherapy. Our selves are not "contained" in inner worlds bounded by our bodies; our bodies are part of our selves—already *with* other selves—and should be understood therefore as a matrix of relationships. Yalom feels that authentic relationships or its group therapy equivalent—cohesiveness—cannot occur except in what he calls the "here-and-now." Because the bonding of the group occurs in the present, any therapeutic facilitation of that bonding process must also be in the here-and-now. The past and future do not have to be *brought into* therapy. They are always and already there, as part of the immediate experiential present. In this sense, the group constitutes a microcosm of society *and* time. That is, the "here" of the group is reflective of the society outside the group, and the "now" of the group is reflective of the past and future. The process works in the reverse if the problem is satisfactorily altered in the group. Any change in relationships or authenticity within the group simultaneously transfers itself outside the group and simultaneously reconstitutes the patient's past (and future). As our present changes, so do the meanings of the past and future, and as our pasts and futures change, so do the meanings of our present.

Family Therapy. Family therapy would seem to give the hermeneutic therapist access to important contextual relations. As Gadamer writes, "Long before we understand ourselves through the process of self-examination, we understand ourselves in a self-evident way in the family, society, and state."[51] A family is its own culture, in a sense, helping to "construct" each family member's "being." This construction is not considered to program the family member (in the linear sense). Whatever influence the family has on the member is dependent upon the context and future "possibility" of *Dasein*. Consequently, no person (or family) is "stuck" in a homeostasis of no "possibility," as some mechanistic systems theorists would claim (see Chapter 8). Furthermore, hermeneutic temporality would challenge the mechanist's definition of a family system—

e.g., "patterns of behavior that are repeated in regular sequences which involve three or more persons."[52] This definition entails universality and reductionism across time. Temporality, by contrast, considers the family system to be wholly present and accessible in the living "now," with all aspects of its "having been" and "might be" implicit therein. The family-as-a-whole is what it is, *as* the family therapist experiences it. The family system is not some transcendent abstraction (e.g., cybernetics) which "spans" the "regular sequence of behavior." The family is an ever-present contextual whole. When a family therapist is part of that whole, opportunities are provided for new meanings and a reconstruction of the collective past.

General Conclusion

We have now outlined two alternatives to the Newtonian framework for time. Each in its own way challenges this framework. First, neither alternative affirms time as an objective entity independent of the events it supposedly contains. Each considers the events themselves (e.g., the whole, *Dasein*) to be the genesis of time. Second, both view "reality" from a holistic perspective, and thereby assert the basic discontinuity of qualitative change. This does not imply that change (in a whole or a meaning) is always cataclysmic. Discontinuous changes are often slight and can be studied through quantitative procedures. Third, both alternatives see change as directional and asymmetrical, with few human processes (if any) running "in reverse." In neither case, however, does this asymmetry lead to a linear metaphor or the sequencing of past, present, and future. Both alternatives view all dimensions of time as co-occurring—whether they are part of a superordinate temporal whole or the implicit temporal context of a "living now." Both consider all dimensions to be vital to the understanding of any one dimension.

Neither alternative agrees with Newton's conception of the universality of time. Similar to the issue of objectivity, Newton's universal laws are considered to be too abstracted from the phenomena they supposedly govern. Each alternative does affirm the essential unity of knowledge (across its own conception of time), but each approaches this unity differently. Many organismic theorists consider a nonlinear type of universal understanding to be possible. Bertalanffy and Chomsky are just two of the more holistically inclined theorists to postulate "deep structures" which do not change across history, despite their nonobjectivity and discontinuity.[53] Most hermeneuticists, however, disagree with this position. Although human meanings often have a unity or "structure" (e.g., a story), these structures depend upon cultures and traditions which are themselves subject to historical change.

Finally, each alternative has its own problems with Newton's reductionism. Each argues for broad and encompassing forms of holism that make reduction to any one element or meaning unproductive, if not unintelligible. Yet, here again, an interesting difference can be seen between theorists representing the two alternatives. This difference appears to center on the subject/object dichotomy. Although organismic holists contend that the observer and observed are holistically related, this relationship does not prevent some holists from considering the "object" to be "outside" the observer. Many holists would argue that the observer and object are "inside" a superordinate structure. However, the observer (as part of the structure) may still be viewed as "subjectively" perceiving "objective" reality. Hermeneutic temporality, on the other hand, denies this arrangement at any level of analysis. The hermeneutical interpreter and interpreted are inextricably bound together, neither being "outside" the other.

Before closing, it should be noted that neither temporal alternative is described sufficiently to begin a school of psychology or instigate a program of research. This is not the reason for their description here (though interesting research is certainly taking place). The purpose of the alternatives is to serve as contrasts to the prevalent assumptions and characteristics of Newton's framework. Without such contrasts, the axiomatic status of this framework renders it almost impervious to thoughtful examination. Linear time would be seen as a widespread but uninteresting and insignificant commonality among explanations—like being under the same sky. With contrasts, however—especially contrasts that psychologists themselves have generated—the prevalence of linear time must be taken more seriously. In a sense, two challengers have emerged to fight for comparable conceptual territory. The extent of their challenge is perhaps in the eye of the beholder, but it is hoped that they will facilitate a healthy skepticism of linear time as well as a systematic examination of its proper place in psychology.

NOTES

Introduction

1. Lovejoy, 1961, p. 75.

2. E.g., Rakover, 1990, p. 44. In this regard, one could imagine two identical (but independent) clocks, one set to chime a little before the other. The sequence of chimes each hour would be perfectly correlated, yet the first would not be considered to cause the second.

3. E.g., Rychlak, 1988, ch. 6.

4. Notable exceptions include Faulconer and Williams (1985), McGrath and Kelly (1986), Rychlak (1981 and 1988), Slife (1981, 1989, and 1991c), and Williams (1990a).

5. E.g., Coveney and Highfield, 1990; Prigogine and Stengers, 1984.

6. E.g., Coveney and Highfield, 1990, pp. 24 and 26. Some theorists may use the term "linear" to signify *only* directionality (or irreversibility), and thus do not confound the characteristics of linear time (as defined in the present book) with time itself. However, a significant number of theorists not only seem to view "linear" and "irreversible" as interchangeable terms, they also consider the irreversibility of time to include distinctly linear properties (see Chapter 1 below). Coveney and Highfield (1990), for example, consider the fixedness of the past (p. 23) and the "bit by bit" flowing of time (p. 24) as part of its irreversibility. The problem is that irreversibility does not require these linear properties.

7. Cf. Morris, 1984, ch. 12.

8. As other sections of this book show, some humanists, phenomenologists, and existentialists view the lived "future" as being the most influential of the temporal dimensions.

9. E.g., Bertalanffy, 1968; Bohm, 1980; Jung, 1960; Lewin, 1936. More specific reference information can be found in this book: Bertalanffy (see Chapter 8), Bohm (see Chapter 10), and Jung and Lewin (see Chapter 3).

10. Pepper, 1970.

11. Bertalanffy, 1968.

12. Jung, 1960.

13. Lewin, 1936.

14. Bohm, 1980.

15. E.g., Coveney and Highfield, 1990; Hawking 1988.

16. Kuhn, 1970; Slife, 1991c.

Chapter 1: Newtonian Time and Psychological Explanation

1. Cf. Faulconer and Williams, 1990.

2. Porter, 1980.

3. Cf. Porter, 1980; Morris, 1984, ch. 1 and 2; Whitrow, 1980, section 2.3.

4. Porter, 1980, p. 13.

5. Whitrow, 1980, section 2.3.

6. Christian doctrine may have been interpreted historically as supportive of a linear view, and thus led to its rise in Western culture. Nevertheless, it is debatable whether Christian doctrine *implies* linear time. Biblical revelation points to the directionality of events, but it does not necessarily point to a linear interpretation of this directionality (see Introduction). The past, for example, is not necessarily primal in Christian theology. The present (e.g., the indwelling of the Holy Spirit) and the future (e.g., Christ's Second Coming) seem to be as important as the past. Moreover, God is often thought to transcend time, and sinners are viewed as being "reborn" and thus transcending their pasts.

7. Mumford, 1934, p. 14.

8. Whitrow, 1980, pp. 58-59.

9. Mumford, 1934, p. 17.

10. Morris, 1984, ch. 3.

11. Not only was cyclical time possible, but Newton also differentiated his "absolute time" from "relative time" (Newton, 1687/1990, p. 8). Moreover, Leibniz opposed Newton's conceptions with his notion of "relational time." From Leibniz's perspective, time does not exist in its own

right independently of events. Time is the successive order of the events themselves (Whitrow, 1980, p. 36–39).

12. As cited in Holton, 1973, p. 5.

13. Cf. Burtt, 1954, ch. 7, sections 3 and 4.

14. This is not to imply that Einstein eliminated time. As Whitrow (1980, section 6.5) has noted, modern notions of "cosmic" time are not incompatible with relativity theory.

15. McGrath and Kelly, 1986, pp. 26 and 30.

16. Burtt, 1954, p. 207.

17. As Burtt (1954) states, "Newton . . . took over without criticism the general view of the physical world and of man's place in it which had developed at the hands of his illustrious predecessors" (p. 231). His most immediate predecessor was Isaac Barrow (1735) who regarded time as "passing with a steady flow" (p. 36). Williams also points to Aristotle as one of the primary philosophical precursors of Newton's view of time (Faulconer and Williams, 1985; Williams, 1990). In addition, David Hume contributed to Newton's legacy. Although Hume was born 69 years after Newton, Hume's explicit fusion of linear time and causality was important to Newton's students.

18. Burtt, 1954, p. 230.

19. Bateson (1978), Leahey (1987), Overton and Reese (1973), Polkinghorne (1983), Rychlak (1981 and 1988), Slife (1981, 1989, and in press).

20. Newton, 1687/1990, p.8.

21. Burtt, 1954, p. 249.

22. Whitrow, 1980, pp. 185–90.

23. Bunge, 1959, pp. 62–64.

24. Newton was a highly religious man whose theology guided much of his scientific work (Burtt, 1954, pp. 256–64). God for Newton was the First Cause of the world, and thus time has a beginning point (unlike cyclical time), and properties akin to a geometric line.

25. This linear characteristic of absolute time may seem in contradiction with Newton's contention that natural processes are "reversible." Newton believed, for example, that celestial mechanics are consistent with his mathematical laws whether they are run "forward" or "backward." This would seem to imply that time itself can run forward and backward, e.g., Coveney and Highfield (1990, p. 30) and McGrath and Kelly (1986,

p. 29). However, Newton made clear distinctions between natural change processes and absolute time. Absolute time flows "without relation to anything external" and is independent of these natural processes. Natural processes may be reversible, but their temporal medium—absolute time—is not. How else could we know, Newton might ask, whether natural processes of change are forward or backward, unless we have some absolute standard by which to judge this directionality? As Newton (1687/1990) described it, "The flowing of absolute time is not liable to any change ... the order of the parts of time is immutable" (pp. 9–10). For the purposes of this book, reference to Newtonian time is a reference to absolute time, which has definite linear properties.

26. Of course, whenever the past is fully understood in a Newtonian framework, the present and future are also considered to be predictable as well.

27. Newton, 1687/1990, p. 8.

28. An interesting exception to this was Newton's conception of gravitation. For him, gravity was a force that acted instantly across the distance between one mass and another (Nicolson, 1980, p. 165). As Bunge (1959, p. 64) notes, the notion of instantaneous physical actions was actually quite prevalent during this period of history.

29. Schrag, 1990, p. 65.

30. Faulconer and Williams, 1985, p. 1180.

31. Ballif and Dibble, 1969, p. 32.

32. How well machines actually embody these characteristics is, of course, open to debate. The order that machines represent does not have to be considered linear in nature. That is, the directionality implicit in mechanisms does not have to be framed in absolute and linear terms (see Introduction). Historically, however, machines and linear time have tended to be highly associated.

33. Whitrow, 1980, p. 33.

34. Ibid. Morris (1984) presents a similar challenge: "Nor can any meanings be attached to the statement that time 'flows equably.' If the flow of time is not uniform, how can one measure its irregularities?" (p. 210).

35. E.g., Burtt, 1954, pp. 256–64.

36. Morris, 1984, p. 209.

37. Burtt, 1954, p. 208.

38. Ariotti, 1975; Harris, 1988, pp. 48–51.

39. E.g., Morris, 1984, ch. 12.

40. Cf. Polkinghorne, 1983, pp. 136–37.

41. E.g., Burtt, 1954, p. 229.

42. Mach, 1959, p. 90.

43. See also DeBroglie, 1949.

44. Morris, 1984, pp. 209–10.

45. Einstein did not believe, however, that causal relations were relative to each observer's inertial frame of reference. He considered the order of cause and effect to be invariant and absolute (Ballif and Dibble, 1969, p. 412).

46. Nicolson, 1980.

47. Wolf, 1981, pp. 83–84.

48. Bohm, 1980, p. 128. Bohm cites other examples of instantaneous change, including empirical experiments (pp. 71–72). Essentially, a molecule of two atoms is disintegrated, resulting in the two atoms flying apart. While the atoms are in flight (and potentially separated by great distances), any attempt to measure the spin of one atom is instantaneously registered in its "sister" atom. No time has occurred in which to allow any "transmission" from one atom to the other, yet the two are somehow instantaneously related.

49. Ornstein, 1972, p. 79.

50. Faulconer and Williams, 1985, p. 1182.

51. McGrath and Kelly, 1986, p. 24.

52. Ornstein, 1972, p. 81.

53. Emde and Harmon, 1987, p. 1.

54. Ibid., p. 3.

55. Kagan, 1984.

56. E.g., Fischer, 1984.

57. Faulconer and Williams, 1985.

58. Cf. Rakover, 1990, ch. 2.

59. Fuller, 1990, ch. 1.

60. Cf. Ashcraft, 1989.

61. Ornstein, 1972, pp. 82–84.

62. Rakover, 1990, ch. 2.

63. McGrath and Kelly, 1986, pp. 128–31.

64. Aveni, 1989, p. 36.

65. Martindale, 1981, p. 3.

66. E.g., Nichols, 1984, p. 421.

67. Dreyfus, 1979, ch. 10.

68. Rychlak, 1988, pp. 47–49; cf. Slife, in press.

69. Rychlak (1988) is a notable exception because he has accused psychological experimenters of an "S-R Bind." That is, an S-R type of linear framework is *imposed onto* a nonlinear style of theorizing to make the latter more "scientific" (in the linear sense). This is perhaps most readily seen in the confounding of stimulus-response *theory* with the independent variable-dependent variable of *method* (Rychlak, 1988, pp. 172–74).

Chapter 2: Developmental Psychology

1. Cf. Kagan, 1980.

2. Cf. Birren and Cunningham, 1985.

3. Lipsitt, 1983, p. 192.

4. Kelly, as quoted in Moss and Susman, 1980, p. 531.

5. For exceptions see Neugarten (1980) and Schaie and Hertzog (1985).

6. Cf. Kagan, 1984; Tomlinson-Keasey, 1985, p. 5.

7. Aries, 1962.

8. Porter, 1980.

9. Mumford, 1934.

10. As Whitrow (1980) observes, nearly all Greek philosophers regarded time as periodic. Proclus, for example, referred to the Great Year, recurring again and again: "It is in that way that Time is unlimited [for] the motion of Time joins the end to the beginning and this an infinite number of times" (p. 26).

11. Aristotle, 1990, p. 303.

12. Kagan, 1984.

13. This industrialization also facilitated the dominance of linear time and mechanistic metaphors (see Chapter 1).

14. Locke, 1689/1990, p. 89.

15. Leahey, 1987, p. 108

16. Mendelson (1980) suggests that evolution was primarily a social theory transformed into a biological theory. In this sense, the theory of evolution was the product of a culture which was itself dominated by assumptions of absolute time.

17. Darwin, 1872.

18. Tomlinson-Keasey, 1985, p. 23.

19. See Chapter 3 below and Rychlak, 1981, Chapter 1.

20. Kagan, 1984.

21. E.g., Baltes and Brim, 1983 and 1984.

22. Schaffer, 1988, p. 91.

23. This implication of time is often forgotten because we rely so much on our memories to understand across-time relationships. However, even memory is a "recording device" with its own limitations and biases.

24. Morris, 1990, p. 336.

25. E.g., Schiamberg and Smith, 1982, pp. 120–21; Stott, 1971.

26. Carlson and LaBarba, 1979.

27. Cf. Tomlinson-Keasey, 1985, ch. 3.

28. Bjorklund, 1988.

29. Klaus and Kennel, 1976 and 1982.

30. Lamb, 1982a.

31. See Schiamberg and Smith, 1982, ch. 5; Tomlinson-Keasey, 1985, pp. 161–69, for reviews.

32. See, e.g., Santrock, 1986.

33. For a discussion of Piaget's more anomalous theorizing, see Chapter 5 below.

34. Main, 1973; Matas et al., 1978; Waters et al., 1979.

35. Belsky and Rovine, 1988.

36. E.g., Craig, 1989.

37. Jensen and Kingston, 1986.

38. Hall, 1982.

39. Baumrind, 1972.

40. Craig, 1989.

41. Rutter, 1989.

42. Williams, 1977.

43. Rutter, 1989; Tomlinson-Keasey, 1985.

44. E.g., Fischer, 1984, p. 96.

45. Tomlinson-Keasey, 1985, p. 262.

46. Cf. Fischer, 1984.

47. Tomlinson-Keasey, 1985, pp. 26, 262.

48. Cf. Santrock, 1986, pp. 20–23.

49. See Rutter (1984) for the different meanings of "discontinuity."

50. Emde and Harmon, 1984, p. 1.

51. Ibid., p. 1.

52. Ibid.

53. *Newsweek*, 18 January, 1982, p. 95.

54. Winick, et al., 1975.

55. Kagan, 1971; Kagan et al., 1978.

56. Kagan, 1984, p. 16.

57. Emphasis added, Emde and Harmon, 1984, p. 1.

58. Rutter, 1984, p. 63.

59. Kagan, 1981, p. 39.

60. Baltes, 1983.

61. Brim and Kagan, 1980, p. 1.

62. Rutter, 1984, p. 42.

63. E.g., Emde and Harmon, 1984; Santrock, 1986, pp. 20–23.

64. Kagan, 1984.

65. Kagan and Moss, 1962.

66. Brim and Kagan, 1980.

67. Thompson and Lamb, 1984, p. 315.

68. Emde and Harmon, 1984, pp. 1–2.

69. Lipsitt, 1983.

70. Rutter et al., 1983.

71. Robins, 1978.

72. Rutter and Giller, 1983.

73. Rutter, 1984.

74. Kagan, 1984, p. 17.

75. Ellis, 1900, p. 250.

76. Brim and Kagan, 1980; Kagan, 1984; Scarr, 1987.

77. Scarr, 1987.

78. Baldwin and Baldwin, 1973.

79. Oppenheim, 1981.

80. Kagan, 1984.

81. Kuhn, 1970.

82. Cf. Brim and Kagan, 1980; Emde and Harmon, 1984.

83. Kagan, 1984; Rutter, 1984.

84. Scarr, 1987.

85. Ibid., p. 65.

86. Emphasis added, Scarr, 1987, p. 66. The emphasis upon concurrent events is that of a more structuralistic outlook.

87. Lamb, 1982b.

88. Schaffer, 1988, p. 91.

Chapter 3: Personality Theory

1. Hall and Lindzey, 1978, p. 4.

2. Ibid., pp. 20–28.

3. Ibid., pp. 22–25.

4. Maddi, 1989, p. 8.

5. Phares, 1984, p. 673.

6. Ryckman, 1989, p. 4.

7. Ewen, 1988, p. 523.

8. Maddi, 1989, p. 12.

9. Ibid.

10. Ross, 1987, p. 160.

11. Ryckman, 1989, p. 87.

12. Ibid.

13. Rychlak, 1981, ch. 1.

14. Leahey, 1987.

15. Other early empiricists such as Hobbes affirmed the independence of the external world (see Leahey, 1987, p. 99).

16. As Kant (1990) noted, "But, though all our knowledge begins with experience, it by no means follows that all arises out of experience" (p. 14).

17. Jones, 1969, pp. 84–90.

18. Ibid.

19. Kant intends neither mysticism nor solipsism. In fact, his intent was to save science from Humean skepticism by showing that knowledge did not depend upon finite or incomplete experience (Jones, 1969). Moving causation into the mind of the scientist saved it from mysticism because all minds supposedly work in this manner. Kant also avoids solipsism. Although the scientist can never know the noumenal realm empirically, we *can* obtain valid information about phenomena, and phenomena are not solely determined by the mind.

20. Kant's "subjectivity" here does not preclude intersubjectivity. Indeed, the fact that we all supposedly possess the same categories of understanding means intersubjectivity is expected.

21. Rychlak, 1981.

22. As Garson (1971) and Porter (1980) note, philosophers seem to divide themselves into two main camps with respect to time: subjectivism and objectivism.

23. Locke, 1689/1990, p. 89.

24. Benjamin, 1966.

25. Porter, 1980, p. 36.

26. E.g., Russell, 1959, p. 241.

27. Kant, 1781/1966, p. 31.

28. Kant did not appear to distinguish between time and linear time (see Introduction). It is generally accepted that Kant's conception of time was similar to Newton's (i.e., linear), except that Kant viewed the characteristics of time as subjective and *a priori* (Benjamin, 1966; Jones, 1969). This leaves open the question of whether time—when defined as change only—is "subjective," i.e., in the phenomenal realm only. Nonetheless, it is clear that he contended the subjectivity of *linear* time, and this aspect of Kant's philosophy is the focus of our present discussion.

29. The concept of linear time for Kant "does not inhere in objects but merely in the subject which intuits them" (as quoted in Whitrow, 1980, p. 51).

30. Rychlak, 1981, pp. 13, 20.

31. As Jones explains (1969 p. 85), Kant believed in *both* a noumenal and a phenomenal self.

32. Ryckman, 1989, p. 82.

33. Ewen, 1988, p. 517.

34. Maddi, 1989, p. 50.

35. Freud, 1909/1957, p. 38.

36. Cameron and Rychlak, 1985, p. 148.

37. Cf. Cameron and Rychlak, 1985; Rychlak, 1981, ch. 1.

38. Freud, 1914/1957b, p. 171.

39. Cf. Cameron and Rychlak, 1985.

40. Freud, 1923/1989, p. 650.

41. Freud, 1920/1990, p. 648.

42. Freud, 1926/1959, p. 97.

43. Freud, 1923/1961, p. 28.

44. Rychlak, 1981, p. 49.

45. Slife, 1987a.

46. Cameron and Rychlak, 1985, p. 34.

47. Emphasis added, Menninger, 1966, p. 92

48. Cameron and Rychlak, 1985, p. 108.

49. Rychlak, 1981, ch. 1; see also Chapter 9 below for a description of four forms of determinism (i.e., the "four causes" of Aristotle).

50. Slife, 1981.

51. Emphasis added, Freud, 1916/1963, p. 67.

52. Freud's teleological determinism is dealt with in more depth in Chapter 6 below.

53. Emphasis added, Freud, 1897/1959, p. 233.

54. Emphasis should be placed on "determinism" here because Freud never felt that one could escape the past entirely. It always formed part of the context of the present.

55. Dollard and Miller, 1950, p. 25.

56. Emphasis added, Miller and Dollard, 1941, p. 59.

57. Dollard and Miller, 1950, p. 122.

58. Cf. Rychlak, 1981, ch. 8.

59. Rychlak, 1981, p. 386.

60. Emphases added, Skinner, 1974, pp. 167–68.

61. Emphasis added, Rogers, 1963, pp. 271–72.

62. Skinner, 1974.

63. Slife, 1981.

64. Bandura, 1986, p. 39.

65. Ibid., p. 16.

66. Ibid., p. 22.

67. Ibid., p. 22.

68. Ibid., ch. 2.

69. The capacity of some structuralists to escape linear time in this manner is also discussed extensively in Chapter 4 below.

70. Bandura, 1986, p. 25.

71. Without inborn personal factors, the environment always precedes and thus controls (via linear causation) any personal factors. In this sense, Kantian inborn factors serve to place the personal factors on par (and thus simultaneous) with the environmental.

72. Emphasis added, Bandura, 1986, p. 368.

73. Rychlak, 1981, ch. 7.

74. Ibid, p. 489.

75. Bandura, 1973, p. 53.

76. Ibid., pp. 39, 368.

77. Ibid., p. 39.

78. Ibid., p. 368.

79. Hall and Lindzey, 1978, p. 686.

80. In this regard, Lewin has been dropped from most modern personality texts.

81. Several nonlinear theorists value some continuity of personality across time, though they consider this the lack of change in a discontinuous system. Allport, for example, valued personality traits but affirmed their "functional autonomy" as in this passage: "The dynamic psychology proposed here regards adult motives as infinitely varied, and self-sustaining, *contemporary* systems, growing out of antecedent systems, but functionally independent of them" (emphasis added, Sahakian, 1974, p. 403).

82. Significant also are Jung's studies of Eastern religions and philosophies which helped him to avoid Newtonian metaphysical assumptions.

83. Lewin, 1936, p. 31.

84. Ibid., p. 30.

85. Ibid., p. 35.

86. Ibid., p. 33.

87. Ibid, pp. 34–35.

88. E.g., Skinner, 1957.

89. Newton, of course, is another positivist asserting theoretical constructs that are unobservable. His notion of absolute time is such a construct.

90. Lewin, 1936, p. 33.

91. Ibid., p. 35.

92. Including Lewin, 1936, p. 32.

93. Cf. Rychlak, 1981, ch. 4.

94. Cf. Jung, 1960, pp. 421–22.

95. Ibid., p. 435.

96. Ibid., p. 436.

97. Einstein, 1961/1990, p. 237.

98. Jung, 1960, p. 438.

99. Bohm, 1980, pp. 71–72; Wolf, 1981, p. 168.

100. Capra, 1982, pp. 80–88.

101. Wolf (1981) attempts one such linear explanation: " 'Something' had to go from the location of the measurement of the first particle to the site of the second . . . [implying] that this 'something' was perfectly able to move at lightspeed or even faster" (p. 168). This speed of travel is only necessary because a linear account of the two particles is assumed to be required.

102. Exceptions would be Rychlak (1981) and Hall and Lindzey (1978).

103. E.g., Ryckman, 1989, p. 89.

Chapter 4: Psychological Method

1. McGrath and Kelly, 1986, p. 3.

2. Campbell and Stanley, 1963, p. 5.

3. Burtt, 1954, p. 224.

4. As Burtt (1954) notes, Newton did not spell out his particular approach to method anywhere in his writings (pp. 220–26). One must examine his experimental practice to discern his experimental method. A particularly good illustration of Newton's method—with its observations across time—is found in Newton's *Principles* (1687/1990, pp. 11–12), and others are scattered throughout his *Optics* (1704/1990).

5. The modern presumption that the principle of causality requires the "principle of antecedence," as Bunge (1959, p. 63) terms it, stems primarily from David Hume (1739/1911), Book 1, part 3, section 2.

6. E.g., Rychlak, 1977, p. 44.

7. Russell, 1919.

8. Ibid., pp. 204–5; Rychlak, 1977, pp. 44–47.

9. Rychlak, 1977, pp. 40–41.

10. This is probably not giving the metaphysical conflict implicit in this methodological confrontation it's just due. Many rationalists have argued that real knowledge is knowledge of the static, eternal, and universal. Because sensory experience is always so changeable, it cannot be real knowledge. The rise of early empiricism (e.g.,with Ockham) can be construed as a rejection of this view of knowledge. The irony is that latter-day empiricists, such as Newton, resurrected the rationalists faith in the unchanging and universal laws supposedly behind our sensory experiences.

11. Rychlak, 1977, p. 47.

12. The need for continuous time in the new mathematics of calculus also played a role in the appeal of temporal universals (Whitrow, 1980, p. 185).

13. Newton, 1687/1990, p. 8.

14. E.g., Leahey, 1987, p. 87.

15. Newton's works are filled with a constant polemic against the use of even "hypotheses" in experimental method (Burtt, 1954, p. 215).

16. McGrath and Kelly, 1986, ch. 6.

17. Ibid., p. 3.

18. Ibid., pp. 149–56.

19. E.g., Kagan, 1984.

20. Emde and Harmon, 1984.

21. Emde and Harmon (1984) describe research strategies that they feel can be used to study the continuities and discontinuities of development. However, these are the less-controlled, "quasi-experimental" variety of research design (cf. Campbell and Stanley, 1966).

22. E.g., Pittendrigh, 1972.

23. McGrath and Kelly, 1986, p. 140.

24. Cf. Campbell and Stanley, 1963.

25. Ibid, p. 34.

26. Ibid, p. 34.

27. McGrath and Kelly, 1986, pp. 152–53.

28. E.g., Fuller, 1990; Gadamer, 1975; Faulconer and Williams, 1985; Polkinghorne, 1983.

29. Skinner, 1974.

30. Cf. Banaji and Crowder, 1989; Dreyfus, 1979.

31. Neisser, 1978.

32. Neisser (1991) now feels that cognitive psychologists have incorporated more "naturalistic" approaches in their study.

33. Neisser, 1991.

34. Dreyfus, 1979, ch. 5 and 6.

35. Gergen, 1985, p. 267.

36. Minuchin, 1974; Bertalanffy, 1968.

37. McGrath and Kelly, 1986, pp. 128–29.

38. Slife, 1981, pp. 33–34.

39. Polkinghorne, 1983, p. 166.

40. Polkinghorne, 1983, ch. 4; Tomlinson-Keasey, 1985, pp. 27–30.

41. Bertalanffy, 1928.

42. Polkinghorne, 1983, ch. 4.

43. Emphasis added, quoted in Polkinghorne, 1983, pp. 139–40.

44. Quoted in Polkinghorne, 1983, p. 143.

45. Emphasis added, Saussure, 1907/1966, p. 114.

46. Piaget derived his structural emphasis from his original training as a biologist.

47. Piaget, 1973, pp. 50–52.

48. Dilthey, 1900/1976; Polkinghorne, 1983, p. 218.

49. Ermath, 1978, p. 303.

50. "Begins" is in quotes so that the reader does not confuse this meaning with "begins" in the absolute sense. In the realm of hermeneutics, time is a lived and experienced time, and not an absolute entity that can be used objectively to judge what comes "first." In any case, "begins" connotes here more of a logical rather than a chronological distinction.

51. Rickman, 1979, p. 130.

52. Some see it, quite inaccurately, as a variation upon introspectionism. See Fuller (1990) and Faulconer and Williams (1990) for distinctions.

53. Hermeneuticists are often criticized for being relativistic, i.e., not having any means of transcending context and making judgments or evaluations. Hermeneutic contextualism, however, is intended to be a rejection of traditional metaphysical absoluteness, not a rejection of transcendence. A type of transcendence is considered available in hermeneutics that makes it possible to avoid relativism. This transcendence does not demand that the researcher get outside linear time to get a larger and truer perspective. Rather, the transcendence of hermeneutics is the "transcendence of the world." No one context exhausts the world—there is always excess. Because of this, every context transcends every other, and we are able to make judgments. We just do not transcend absolutely. This type of transcendence is considered possible precisely because there is no linear time—precisely because the present manifests the givenness of the past and the openness of the future (cf. Faulconer and Williams, 1985).

54. This is true no matter how small the interval (or separation) between temporal dimensions. Although time is a line, the present is still conceived as a point on the line separating the past and future. Temporal dimensions are thus sequential. Past events may cause present events, but the past is by no means "present" in the absolute temporal framework of Newton.

55. Faulconer and Williams, 1985; Faulconer and Williams, 1990; Heidegger, 1962.

56. Gadamer, 1975.

57. See also "Coherence Theory of Truth" (Rychlak, 1988, pp. 45–46).

58. Faulconer and Williams, 1985.

59. See Packer (1985) for an excellent example of hermeneutic research.

60. Cf., Kuhn, 1970.

61. Burtt, 1954, p. 229.

Chapter 5: Cognitive Psychology

1. Cf. Ashcraft, 1989, p. 21; Ellis and Hunt, 1989, p. 4; Reynolds and Flagg, 1977, p. 4.

2. E.g., Wessells, 1982. Much of our current interpretation of Aristotle stems from our "modernist" perspective (cf. Mueller-Volmer, 1990). For example, modern distinctions between subject and object may have colored

our understanding of ancient Greek philosophy, where it is debatable that there were such distinctions.

3. Hawking, 1988, p. 18.

4. Williams, 1990a.

5. Other principles of association for Aristotle include similarity and contrast. The fact that cognitive models rely less upon these principles than those associated with linear time is further evidence of the basic linearity of cognitive psychology.

6. Robinson, 1976, p. 84.

7. Wessells, 1982, p. 16.

8. The "learning curve" is itself a product of linear assumptions (see Chapter 9 below). The data of a learning curve only occur as separated points along this curving line. Because time is considered to be continuous, intervening learning states are also presumed to be continuous. The data points are thus linked as though they occur in one continuous flow, but in actuality no such "continuous flow" is ever observed.

9. Noble, 1952.

10. Archer, 1960.

11. Wessells, 1982, p. 20.

12. Searle, 1985.

13. Rychlak, 1981, p. 72.

14. E.g., Ellis and Hunt, 1989.

15. E.g., Wessells, 1982, p. 14.

16. Rychlak, 1981, ch. 1; see also Chapter 3 above.

17. As Kant (1781/1966) observes, "Time is not an empirical concept deduced from any experience, for neither coexistence nor succession would enter into our perception, if the representation of time were not given *a priori*. Only when this representation *a priori* is given, can we imagine that certain things happen at the same time (simultaneously) or at different times (successively) " (p. 29).

18. For an exception, see Lewin (1936) and Chapter 3 above.

19. Bartlett, 1932, p. 213.

20. Ashcraft, 1989, pp. 304–11.

21. See also Matlin, 1983, pp. 163–66.

22. Rychlak, 1981, ch. 11.

23. Bartlett, 1932, p. 201.

24. As noted in the Introduction, sequentiality is not the exclusive province of linear viewpoints. Cognitive processes can evidence temporal order without also being linear in nature. In this particular case of sequence—i.e., input → output—an important feature of the explanation is the linear arrow, indicating the primacy of the past.

25. This is not to say that all rationalist accounts avoid Newtonian explanatory characteristics or Newtonian metaphysics. Kant's philosophy, for example, is fundamentally universalist, as Heidegger (1962) noted.

26. E.g., Ellis and Hunt, 1989; Ashcraft, 1989.

27. Ellis and Hunt, 1989, pp. 36, 60–61, 84, 113–14 for summary.

28. Wessells, 1982, p. 130.

29. Craik and Lockhart, 1972.

30. Craik and Watkins, 1973.

31. Wessells, 1982, p. 132.

32. Craik and Lockhart, 1972.

33. E.g., Ellis and Hunt, 1989, ch. 5.

34. E.g., Matlin, 1983, pp. 25–26.

35. Emphasis added, Ashcraft, 1989, p. 64.

36. Emphasis added, Ibid., 1989, p. 73.

37. Kintsch, 1977.

38. Loftus, 1979.

39. Perrig and Kintsch, 1985.

40. Bransford and Franks, 1971 and 1972.

41. Reiser et al., 1985.

42. As Ashcraft (1989) puts it, the overriding theme of these cognitive approaches is "how remembering is affected by *existing* knowledge" (emphasis added, p. 305).

43. Emphasis added, Ashcraft, 1989, p. 338.

44. Rychlak, 1979, ch. 1.

45. Abbot et al., 1985; Reiser et al., 1985; Schank and Abelson, 1977.

46. Schank and Abelson, 1977.

47. Reiser et al., 1985.

48. Ashcraft, 1989, p. 478.

49. Compare this with Kagan's assertion (in Chapter 2 above) that he often does not see such "seamless transitions" in his own developmental data. In fact, he speculates that many of the intermediate stages considered necessary for continuous transition in language development (e.g., two-word utterances) do not exist for many children.

50. Chomsky, 1966 and 1968.

51. Cf. Anderson, 1980; Ellis and Hunt, 1989; Matlin, 1983; Wessells, 1982. Some texts differentiate Chomsky from Skinner's brand of empiricism, but this differentiation typically does not involve *cognitive* distinctions between rationalism and empiricism (e.g., Ashcraft, 1989, pp. 26–28; Matlin, 1983, p. 4).

52. E.g., Ashcraft, 1989, pp. 498–503.

53. Ibid, pp. 504–5.

54. Ibid., p. 505.

55. Emphasis added, Ibid., p. 523.

56. E.g., "insight learning," Kohler, 1927.

57. E.g., Newell and Simon, 1972; Simon, 1979.

58. E.g., Melkman, 1988.

59. Ibid., p. ix.

60. Emphases added, Evans, 1973, p. 20.

61. Piaget, 1967, p. 10.

62. In this sense, assimilation to a preexisting biological structure comes before accommodation to the environment (Rychlak, 1981). This does not imply a linear thesis, because the environment and innate structures are simultaneous in assimilation.

63. Some cognitive researchers make the distinction between serial and parallel processing (e.g., Salthouse, 1984). Serial processing connotes a series of processes occurring one at a time, whereas parallel processing connotes two or more processes (each happening across time) occurring simultaneously. This simultaneity of processing, however, is still thought to occur across linear time and often contains no holistic transcendence, despite the simultaneity of relationships.

64. Piaget, 1970, p. 9.

65. Piaget, 1970, pp. 118–19.

66. Emphasis added, Melkman, 1988, p. 2.

67. Historians can also be viewed as reconstructing the past. An example of the present reconstrual of history—from the culture of psychology itself—is the widespread conception that John Watson is the "father of behaviorism." This conception usually implies that his ideas not only preceded other ideas like it but that his ideas produced or stimulated the behavioral ideas following it. As Samelson (1981) notes, however, published responses to Watson's behavioristic manifesto were few in number and muted in tone. Watson was not viewed at the time as asserting anything particularly new. He merely rode the crest of a behavioral movement that was already well underway. As the historian, Thomas Leahey (1987) puts it, "[Watson's manifesto] simply marks the moment when behavioralism became ascendant and created for later 'behavioralists' a useful 'myth of origin'" (p. 306). In other words, Watson's paper was not considered "the beginnings of a revolution" until much later in a retrospective reconstruction of the events.

68. Cf. Lewis and Williams, 1989.

69. Ellis and Hunt, 1989, ch. 12.

70. Bower, 1986.

71. Cf. Ellis and Hunt, 1989; Kuiken, 1991.

72. Rychlak, 1988, ch. 9 and 10; Slife, 1987a; Slife and Rychlak, 1981 and 1982; Slife et al., 1984.

73. Rychlak, 1988.

74. Ibid., p. 275.

75. Slife, 1981 and 1987a.

76. Some linear theorists also claim that future-related constructs, such as goals and anticipations, are in the "present" (e.g, Bandura, 1986). Because causes must precede effects, the only way for a goal to cause behavior in a linear model is to bring the goal into the present *before* the behavior it supposedly causes. This is a linear (and not a teleological) explanation for two reasons. First, most telic theorists view not only the "cause" but also the "effect" as in the present (i.e., simultaneously). These teleologists would permit the possible (discontinuous) shift of a goal in the next "moment" (before a linear "effect" of the goal could have occurred). Unlike the linear theorist, the goal only affects the behavior with which it is contemporaneous. Second, the teleologist considers the final cause (goal

or plan) to be an original contribution to behavior. The linear theorist would hold that such goals and anticipations are ultimately determined by chains of causation from the past.

77. Rychlak, 1988, pp. 243–45.

78. Ibid, p. 278.

79. Dreyfus, 1979, p. 231.

80. Obviously, cognitions always have content which is related to the environment. However, this does not mean that the environment has to *produce* the cognitions. As Kant (1781/1966) noted, experience is necessary for all knowledge, but knowledge does not arise out of experience.

81. Slife, 1987a.

82. Rychlak, 1988, ch. 9 and 10.

83. Rychlak, 1979, ch. 1.

84. As two prominent cognitive researchers, Alan Newell and Herbert Simon, have noted, "The programmed computer and human problem solver are both species belonging to the same genus 'Information Processing System'" (Weizenbaum, 1976, p. 169).

85. E.g., Minsky, 1986.

86. Rose, 1985, p. 47.

87. Dreyfus, 1979; Dreyfus and Dreyfus, 1986.

88. Dreyfus and Dreyfus, 1986, p. xii.

89. Dreyfus, 1979, p. 231.

90. Oettinger, 1979, p. xiii.

91. Winograd and Flores, 1987.

92. Ibid.

93. This places Dreyfus outside the traditional rationalist camps of Kant and Descartes, as he explicitly notes. Although Dreyfus's existential views clearly make him more rationalist than empiricist, his challenge of universality puts him at odds with many rationalists. Kant, for instance, asserted the existence of universal categories of understanding that are independent of temporal contexts (see Chapter 3 above).

94. Dreyfus, 1979, p. 240.

95. Merleau-Ponty, 1962, p. 4.

96. Winograd and Flores, 1987.

97. Dreyfus, 1979, p. 241.

98. Merleau-Ponty (1962) speaks of a "synergistic system" with "a ready-made system of equivalents and transpositions from one sense to another" (pp. 234–5).

Chapter 6: Individual Therapy

1. Becvar and Becvar, 1988, p. 23.

2. As Leahey (1987) notes, "The concept [of the individual] was invented during the Middle Ages" (pp. 68–69). For the most part, medieval thinkers had no *conception* of the individual as an important object of concern or study. This was because the Neoplatonic zeitgeist dictated that the human intellect know only universals, in this case humanness rather than individuals. Not until the nineteenth century do we find a systematic interest in individual differences. Of course, other factors than linear time have been associated with this change in perspective, including ethics, capitalism, and Protestantism.

3. Reeves, 1982, p. 119.

4. Cf. Garfield and Kurz, 1976.

5. Theories of development and cognition could be included here. They have also served as theoretical foundations for many psychotherapeutic approaches. Nevertheless, they have not traditionally been as closely tied to individual psychotherapy as personality conceptions (cf. Rychlak, 1981).

6. Cameron and Rychlak, 1985.

7. Rychlak, 1981, ch. 4; Cameron and Rychlak, 1985.

8. One of the reasons that the telic elements of Freud have been overlooked is the mistranslation and misconstrual of his concept of "instinct." As Bettelheim (1983, pp. 103–12) has suggested, Freud's meaning is not the typical American meaning of a passive, hard-wired and genetic characteristic. Instinct is better translated as impulse, wish, or even intention—any one of which better connotes the telic side of Freud's theorizing.

9. Emphasis added, Freud, 1916/1963, pp. 58–59.

10. Rychlak, 1981, pp. 300–4.

11. A frequent misconception of final causes is that linear causes (across time) determine them. Certainly the meaning of the past can form a context for final causes. However, goals and intentions are thought to initiate patterns of behaviors that are themselves uncaused by the objective past (cf. Rychlak, 1981, introductory ch.). The fact and frequency of

this misconception is another indicator of the pervasive influence of Newtonian linear time (and hence, linear causation).

12. Rychlak, 1981, p. 416.

13. Dollard and Miller, 1950, p. 8.

14. Jung, 1960a, 405–6.

15. E.g., Jung, 1960b, p. 190.

16. Jung, 1959, p. 173.

17. Emphasis added, Jung 1954, p. 34.

18. As cited in Kaufmann, 1989, p. 133.

19. Ibid., p. 134.

20. Adler also favored a teleological understanding of human evolution. Contrary to Darwin's more mechanistic version, Adler postulated that humans behaved for the sake of telic goals in accounting for evolutionary "movement."

21. Adler, 1929/1974, p. 85.

22. Adler, 1958, p. 26.

23. Adler, 1929/1974, p. 85.

24. Emphasis added, Arlow, 1989, p. 29.

25. Ibid.

26. Arlow's (1989) article is part of a standard and widely used text on psychotherapy. The editors of that text—Raymond Corsini and Danny Wedding (1989)—claim that "chapters are carefully selected to reflect changes in a constantly evolving field," and that "leading figures in psychotherapy are recruited as authors" (p. ix).

27. Emphasis added, Arlow, 1989, p. 39.

28. Ibid.

29. E.g., Blanck and Blanck, 1986.

30. E.g., Kohut, 1978.

31. E.g., Mahler, 1968.

32. Freud, 1933, p. 75.

33. Kohut, 1978, p. 480.

34. Ibid., p. 481.

35. Crider et al., 1989, p. 480.

36. Wilson, 1989, p. 242.

37. Ibid., p. 243.

38. Ibid., p. 242.

39. As discussed in Chapter 5 above, many psychologists consider cognitive explanations to be revolutionary in their move away from traditional behavioral notions. Although this is probably true in many ways, there is no evidence in mainstream cognitive theorizing that the basic Newtonian framework for time has been substantially altered.

40. Wilson, 1989, p. 243.

41. Ibid, p. 264.

42. Ibid, p. 250.

43. Ibid., p. 254.

44. E.g., Beck and Weishaar, 1989, p. 290; Ellis, 1989, p. 210.

45. Beck and Weishaar, 1989, p. 294.

46. Ibid, p. 293.

47. Ibid.

48. Ellis, 1989.

49. Some therapy approaches advocated by Rogers and Perls are also explored in Chapter 7 below.

50. Binswanger, 1956, p. 144.

51. May and Yalom, 1989, p. 363.

52. Ibid., p. 394.

53. Yalom, 1980.

54. Emphasis added, Ibid., p. 10.

55. Emphasis added, Ibid., p. 10-11.

56. Emphasis added, May and Yalom, 1989, p. 382.

57. Emphasis added, Yalom, 1980, p. 29.

58. Slife and Barnard, 1988.

59. As cited in Binswanger, 1958, p. 196.

60. Emphasis added, May and Yalom, 1989, p. 366.

61. Ibid., p. 368.

62. Cf. Loftus, 1979.

63. Cf. Arcaya, 1991.

64. Cited in May et al., 1958, p. 67.

65. Ibid., p. 67.

66. Ibid, p. 66.

67. May and Yalom, 1989, p. 369.

68. May, 1958, p. 72.

69. Goldstein, 1939, p. 30.

70. May and Yalom, 1989, p. 369.

71. Sartre, 1947, p. 15.

72. May and Yalom, 1989, p. 372.

73. Cf. Slife, 1991b.

74. Yalom, 1980, ch. 1.

75. May and Yalom, 1989, p. 375.

76. Ibid., p. 388, cf. Yalom, 1980, ch. 2.

Chapter 7: Group Therapy

1. Vander Kolk, 1985, p. 1.

2. Cf. Slife, 1988.

3. Slife, 1991a; Slife and Lanyon, 1991.

4. Steiner, 1974.

5. Forsyth, 1990, p. 14.

6. Durkheim, 1897/1966.

7. Steiner, 1974; see also Forsyth, 1990, pp. 16–17.

8. Steiner, 1974, p. 96.

9. Lakin, 1985, p. 15.

10. Forsyth, 1990, p. 17.

11. Allport, 1924, p. 8.

12. Ibid., p. 5.

13. Lakin, 1985, p. 12.

14. Cf. Corey, 1990, p. 152; Vander Kolk, 1985, p. 27.

15. See Chapters 3 and 6 above for further information about Freud's "mixed" model of linear and nonlinear characteristics.

16. Forsyth, 1990, p. 463.

17. Wolf, 1963.

18. Smith, 1980, p. 7.

19. E.g., Yalom, 1985, pp. 186–92.

20. Yalom, 1985, p. 194.

21. The demise of Tavistock groups (and group-as-a-whole approaches in general) may be greatly exaggerated. My own informal survey of the 1990 convention of the American Group Psychotherapy Association indicated much interest in and considerable clinical practice of group-as-a-whole techniques.

22. Forsyth, 1990, p. 110.

23. Rudestam, 1985, p. 2.

24. Most self-help groups (e.g., Alcoholics Anonymous) would also fit within this tradition. Although it is rare for a formally trained "expert" to be present in such groups, their purpose is to pass on the lore or educate the members. In this sense, the experienced members of the group assume the expert role, and universalist assumptions about the group are retained.

25. Lewin, 1951, p. 228.

26. Rogers, 1970.

27. Corey, 1990, p. 255.

28. No research has been conducted to reveal the frequency with which these three forms of group therapy are practiced. The assertion that the psychoanalytic and behavioral forms are the more dominant stems from my own experiences (Slife, 1991a) and comments such as those of Corey (1990), "Psychoanalytic theory has influenced most of the other models of group work" (p. 152).

29. Corey, 1990, pp. 153–54.

30. Other advantages are sometimes listed (e.g., Vander Kolk, 1985, pp. 36–37), but the frame of reference is still individualistic.

31. Tuttman, 1986.

32. E.g., Locke, 1961.

33. Lakin, 1985, p. 19.

34. Forsyth, 1990, p. 463.

35. E.g., Wolf & Kutash, 1986.

36. Wolf, 1983.

37. Mahler, 1968.

38. One exception to this is the Adlerian group. Adler (1969) stressed the telic future, though his group therapy was largely directive and individually focused (Corey, 1990, p. 201).

39. E.g., Leahey, 1987, ch. 10.

40. Corey, 1990, p. 383.

41. Rose, 1980.

42. E.g., Kuehnel and Liberman, 1986.

43. Meichenbaum, 1986.

44. Corey, 1990, p. 413.

45. Ibid., p. 383.

46. Ibid., p. 255.

47. Yalom, 1975, p. 19.

48. Yalom, 1985, pp. 48–49.

49. Perls, 1973.

50. E.g., Zinker, 1978.

51. Yalom, 1985, p. 135.

52. Slife and Lanyon, 1991.

53. Yalom and others consider the "future" important to therapy as well. However, this more telic emphasis does not necessarily involve linear time assumptions. This particular future takes place in the present as a present anticipation or expectation. As Yalom (1980) puts it, it is a "future-becoming-present" (p. 11).

54. Yalom, 1985, p. 28.

55. Slife and Lanyon, 1991.

56. Lewin, 1936.

57. Yalom, 1985, p. 182.

58. Slife and Lanyon, 1991; Yalom, 1985, p. 143.

59. Yalom, 1985, p. 30.

60. Yalom, 1983, p. 46.

61. Perls, 1973; Rogers, 1970; Yalom, 1985.

62. Van Inwagen, 1983, ch. 1.

63. Yalom, 1985, p. 185.

64. Yalom, 1985.

65. Rogers, 1970.

66. Mullan, 1979.

67. Corey, 1990, p. 332.

Chapter 8: Family Therapy

1. E.g., Nichols, 1984, pp. 80–83.

2. This is not to say that individuals cannot be treated with an appreciation for their holistic context. As used here, the term "individualistic" implies reductionism to the person without such an appreciation.

3. E.g., Goldenberg and Goldenberg, 1991, ch. 1.

4. Slife, 1988.

5. Bateson, 1972, p. 475.

6. Becvar and Becvar, 1988, p. 20.

7. Becvar and Becvar, 1988, pp. 134, 237.

8. Bertalanffy, 1967, pp. 65–69.

9. Wiener, 1948, p. 14.

10. Ibid., p. 84.

11. Mechanistic sequences can be understood apart from the Newtonian characteristics typically attributed to them. That is, the specific order of their change processes is not prima facie evidence for linearity (giving the past primacy) or continuity (assuming that change is only smooth). See Chapter 10 below for alternative ways of interpreting the order of mechanistic systems.

12. Goldenberg and Goldenberg, 1991, ch. 1.

13. Bateson et al., 1956.

14. Wiener, 1948; Von Neumann, 1951.

15. E.g., Becvar and Becvar, 1988, p. 38; Goldenberg and Goldenberg, 1991, p. 34.

16. Bertalanffy, 1968, p. 68.

17. Goldenberg and Goldenberg, 1991, p. 34.

18. Becvar and Becvar, 1988, p. 38.

19. Bertalanffy, 1967, p. 65.

20. Ibid., p. 70.

21. Bertalanffy, 1981, p. 116.

22. Ibid.

23. Bertalanffy, 1968, ch. 2.

24. Bertalanffy, 1967, p. 67.

25. Ibid., pp. 88–93.

26. E.g., Keeney, 1983, p. 6.

27. Becvar and Becvar, 1988, p. 24.

28. Nichols, 1984, p. 56.

29. Becvar and Becvar, 1988, p. 134.

30. Nichols, 1984, p. 183.

31. Becvar and Becvar, 1988, p. 146.

32. Ibid., p. 145.

33. Ibid., p. 238.

34. Even the terminology of "response" here puts the therapy in a sequential frame of reference because it implies that somewhere back in time the response was caused by a stimulus, the child.

35. Becvar and Becvar, 1988, p. 246.

36. Emphasis added, Jacobson and Margolin, 1979, p. 13.

37. Certainly, no intentional contribution to the system is possible because all family members (or system components) are determined by cause/effect chains from the past. Intentionality typically implies an ability to behave in a manner that is *either* consistent or inconsistent with the past.

38. Bertalanffy, 1981, ch. 9.

39. Watzlawick et al., 1974, p. 9.

40. Haley, 1976, p. ix.

41. Ibid., p. 82.

42. Selvini-Palazoli et al., 1978, p. 3.

43. Ibid., p. 8.

44. Goldenberg and Goldenberg, 1991, p. 184.

45. Selvini-Palazzoli et al., 1978, p. 21.

46. Nichols, 1984, p. 431.

47. Watzlawick et al., 1967, p. 13.

48. Goldenberg and Goldenberg, 1991, p. 188.

49. Nichols, 1984, p. 407.

50. E.g., Goldenberg and Goldenberg, 1991, ch 9.

51. Bertalanffy, 1967, p. 90.

52. Goldenberg and Goldenberg, 1991, pp. 197–201; MacKinnon, 1983.

53. E.g., Tomm, 1984.

54. E.g., Nichols, 1984, p. 398; Watzlawick et al., 1967.

55. Minuchin, 1981, p. 13.

56. Cited in Minuchin, 1981, p. 13.

57. Minuchin, 1981, p. 13.

58. Emphasis added, Ibid., p. 13.

59. Ibid., p. 14.

60. E.g., Becvar and Becvar, 1988, pp. 75–76.

61. Cf. Minuchin, 1981, p. 21.

62. Nichols, 1984, p. 262.

63. Whitaker, 1976, p. 154.

64. Kempler, 1968.

65. Nichols, 1984, p. 265.

66. Binswanger, 1967; Boss, 1963.

67. Heidegger, 1962; Boss, 1963.

68. Nichols, 1984, p. 265.

69. Kempler, 1965; Nichols, 1984, p. 282.

70. Nichols, 1984, p. 268.

71. Whitaker, 1976, p. 183.

72. Ibid., p. 192.

73. Becvar and Becvar, 1988, p. 61.

74. Varela and Johnson, 1976, p. 30.

75. Becvar and Becvar, 1988, p. 76.

76. Keeney, 1983, p. 142.

77. Maturana, 1978, p. 61.

78. E.g., Becvar and Becvar, 1988, p. 77.

79. Ibid., p. 80.

80. Emphasis added, Ibid., p. 77.

81. As illustrated by Becvar and Becvar, 1988, p. 76.

82. Ibid., p. 80.

83. Emphasis added, Ibid., p. 81.

Chapter 9: General Temporal Themes of Explanation

1. Kuhn, 1970; see also Introduction above.

2. Cf. Leahey, 1987, ch. 14.

3. This Newtonian theme (and the one above) is also advocated by Hempel (1965) in his deductive-nomological model of explanation. The per-

vasiveness of absolute linear time is partly why Hempel's model is so influential in psychology (Rakover, 1990, ch. 2).

4. Those psychotherapists who consider themselves to be "eclectic" are not necessarily exempt from this Newtonian theme. Eclecticism can mean that the therapist has eclectically selected the "best" *universal* explanations from the various theoretical orientations available. The fact that the therapist views the explanation as the best collection of explanations does not necessarily change the universalist nature of the explanations. See Slife (1987b).

5. Tolman, 1967, p. 414.

6. Van Inwagen, 1983, ch. 5.

7. This is not to say that the only reason for the subject/object separation is their sequentiality. Indeed, this separation probably preceded the rise of linear time in the history of ideas. The point here is that the modern sequencing of processes, including processes of subject and object, facilitates, if not leads to, subject/object separation.

8. As we shall see in the "anomalies" section, some rationalistic explanations consider the subject and object to occur simultaneously (e.g., Piaget in ch. 5). Still other rationalistic explanations dissolve the subject/object dichotomy altogether (see Chapter 4 above).

9. Many cultures, such as the Hopi Indians and Trobriand Islanders, have no tense in their languages, and do not view events as occurring in causal sequences (Lee, 1950; Porter, 1980). For the Trobriand Islander, events and objects are wholes that do not *become*. If a *taytu* (a type of yam) is overripe, then it is no longer a *taytu*. It is a new entity—a *yowana*—with different characteristics and properties. In other words, Trobrianders do not connect temporal events with lines. They do not consider the *taytu* (then) as continuously linked with an over-ripe *taytu* (now). Their world is one of patterned wholes changing *dis*continuously into other patterned wholes, with separate emergent properties. There is no distinction between past and present and no arrangement of events into causal chains or means/ends relationships (Lee, 1950). Indeed, Trobrianders often do not employ linearity in their view of space. When Western thinkers see a "line" of trees or a "circle" of stones, we assume the presence of a connecting line which is not actually visible. We do this metaphorically when we follow a "line" of reasoning, a "course" of action, or a "span" of time. Trobrianders, however, do not. To Western eyes, their huts are arranged in circles, but the Trobrianders literally never see these circles. In fact, they refer to their village with the same word they use to refer to an unpatterned rash. To them, their huts are best described as an "aggregate of bumps." It may be difficult to understand their seeming blindness to a particular linear organization such as a circle of huts. However, it is important to remember

that these circles are not actually present. They are imposed *onto* reality in much the same way as the Westerner imposes temporal and causal linearity onto the reality of world events.

10. This often leads to the criticism that such contextual approaches are relativisms, yet this stems from the presumed dichotomy between absolute and relative explanations—i.e., if the explanation is not absolutistic, then it must be relativistic. As some hermeneuticists have noted (e.g., Faulconer and Williams, 1985), there are alternatives to this dichotomous thinking (see Chapter 10 below).

11. It is important to note that the lack of formal theorizing (e.g., eclecticism) does not necessarily exempt a psychotherapist from universal theorizing. The therapist can hold universal concepts of treatment (e.g., diagnoses) without identifying with a formal theoretical perspective.

12. They are not linear causes of the linear *future* either. They exist in the "now" as part of the present context determining present attitudes and behaviors. Goals and purposes can be discontinuously switched in the next instant, along with accompanying attitudes and behaviors.

13. Such wholes are actually quite common, as illustrated in the following: "Waiting for her husband, she found herself to be quite angry. However, this anger was completely changed (and perhaps not even remembered) after news of his traffic accident."

14. The term "consolidate" is used here to contrast Heidegger with Newtonian conceptions. However, Heidegger does not just consolidate subject and object. Just as the whole transcends the sum of its parts, so *Dasein* leaves the realm of subject and object language altogether.

15. Of course, there are other grounds for the separation of individuals, e.g., we are separate protoplasms. Anomalous theorists are not questioning the existence of physical individuals; they are questioning the linear *conception* of these physical entities.

16. Rakover, 1990, pp. 32–38.

17. Bunge, 1959, p. 63; cf. Bunge, 1963, p. 189.

18. Bunge, 1959, p. 63; see also Brand, 1976, ch. 1.

19. Emphasis added, Bunge, 1963, p. 189.

20. Bunge, 1959, p. 63; see also Brand, 1976, pp. 8–9; Rakover, 1990, p. 37.

21. Brand, 1976; Bunge, 1959; Overton and Reese, 1973; Rakover, 1990; Rychlak, 1981 and 1988; Slife, 1981 and 1987a.

22. The "circular causation" of family therapy (Nichols, 1984, pp. 127–132) is excluded as an alternate form of causation here. As the analysis in Chapter 8 above demonstrates, this form of causation is a variation on linear (and efficient) causal themes.

23. Cf. Slife, 1981.

24. Rychlak, 1988, ch. 1; Bunge, 1959, pp. 62–64.

25. E.g., Rakover, 1990, pp. 45, 64.

26. Cf. Rychlak, 1988, ch. 6; Slife, 1981 and 1987a.

27. Consider also the telic explanations of Freud and Adler (in Chapter 6) and Rychlak (in Chapter 5).

28. Consider also Jung's synchronicity, Lewin's momentary sections, and Dreyfus's figure-ground relationship. Minuchin, a family therapist, could also be viewed as a formal causal theorist.

29. Aspects of Freud's and Jung's theories clearly show material causal emphases (cf. Rychlak, 1981, ch. 1 and 3).

30. Burtt, 1954, pp. 231–9.

31. Bunge, 1959, pp. 17-21; 1963, ch. 11. See also Bohm, 1980; Rychlak, 1988, ch. 4 and 5; Slife, 1981.

32. Cf. Aristotle, 1990, p. 278; Rychlak, 1981, p. 269.

33. Quantitative change assumes that two observations placed anywhere in the "line" of change are sufficient to discern its trends. For qualitative change, however, the positioning of observation is crucial.

34. Piaget, 1972, p. 10.

35. Cf. Piaget, 1972, ch. 3. Piaget notes, however, that there is also an "integrative characteristic" to stage structures. That is, "the structures constructed at a given age become an integral part of the structures of the following age" (p. 51). This does not preclude discontinuity in the sense meant here. Old structures can be incorporated in new stages of development such that they have new (and discontinuous) meaning.

36. Kagan (1980) considers Piaget to be a "continuous" theorist. Nevertheless, as noted in Chapter 2 above, Kagan's use of the term "continuous" differs from its use in this context. Kagan seems to mean predictable from one stage to the next. Although Piaget did assert that the stages were predictable from one stage to another—as a whole is changed to another whole—this predictability is more discontinuous than continuous in the sense used here.

Chapter 10: Alternative Assumptions of Time

1. Bohm, 1980, p. 173.

2. Polkinghorne, 1983, p. 135.

3. Bertalanffy, 1968, p. 37.

4. Bohm, 1980, p. 173.

5. Ibid.

6. Ibid., p. 175.

7. Newton, 1687/1990, p. 8.

8. Becvar and Becvar, 1988, pp. 278–81.

9. Heisenberg's uncertainty principle is a manifestation of this in physics (Heisenberg, 1958).

10. This does not preclude the possible reversibility of certain natural processes, as Newton notes (see Introduction). Although these processes can be theoretically "reversed," they cannot affect the directionality of absolute time because this directionality is independent of the events occurring in time. According to Newton, the very fact that we know some natural processes are "reversed" is due to the objective standard (and objective forward movement) of absolute time.

11. The implication is that if all such change stops (including all clock and cognitive change), then time itself stops.

12. Similarly, cause-and-effect relations do not take place in a reverse order, but again, this is not due to the absolutivity of linear sequence. Rather, the very definitions of cause and effect give existential precedence to the cause. Effects cannot (logically) exist unless (rather than "until") their causes do. No primacy of the past is necessary here because simultaneity of cause and effect is possible and indeed expected from an organismic perspective.

13. Interestingly, because clock- and calendar-time have become reified as *the* time, natural processes are often viewed as temporally incorrect. For example, spring can arrive "late," and darkness can occur "earlier" (relative to a supposedly real or absolute time).

14. Polkinghorne, 1983, pp. 136–38.

15. Saussure (1907/1966) made a similar point concerning linguistic concepts. As he put it, "Concepts are purely differential, not positively defined by their content but negatively defined by their relations with other terms of the system" (p. 117).

16. Cf. Bohm, 1980.

17. Ibid.

18. Polkinghorne, 1983, p. 138.

19. Bohm, 1980, p. 213.

20. The notion of second-order change in family therapy seems to parallel discontinuous change in many respects. As Watzlawick and his colleagues (1974) note, "second-order change usually appears weird, unexpected, and uncommonsensical" (p. 83). Could the "unexpected" property of such change be due, at least in part, to its discontinuous and, hence, unpredictable nature in the linear (and "commonsensical") sense?

21. E.g., Faulconer and Williams, 1985; Fuller, 1990; Heidegger, 1962; Merleau-Ponty, 1962; Ricoeur, 1985.

22. Heidegger disliked the "existentialist" label for his own works. Nonetheless, there is little doubt that he has influenced and was influenced by many existentialists (cf. Chessick, 1986).

23. Packer, 1985.

24. Kockelmans, 1965, p. 12.

25. Williams, 1990b.

26. Packer, 1985.

27. The unready-to-hand mode is not discussed here for lack of space. This mode is entered when we encounter some problem in our ready-to-hand, practical activity (Packer, 1985). The problem brings our attention to the particular problematic piece of our activity (e.g., the car does not start) so that a reduction of the patterned activity occurs. This reduction, however, is never context-free. The attention to a particular part of the whole is always in reference to the whole activity itself (e.g., frustration because the car can not take you to the post office).

28. Heidegger, 1927/1962, p. 99.

29. Kockelmans, 1965, p. 81.

30. Newton, 1687/1990, p. 8.

31. As cited in Faulconer and Williams, 1990, p. 48.

32. Fuller, 1990, p. 239.

33. Faulconer, 1990.

34. The recognition of a present-at-hand mode is the recognition that spatialized theories of time are not only possible but perhaps even

important to the culture that has devised them. This does not mean, however, that such spatialized theories are fundamental for understanding temporality.

35. As another example, "I *saw* someone come through here (past), who is *now* walking ahead of me (present) and *will* reach our common destination before me (future)." All temporal dimensions are encompassed in the lived present.

36. Cf. Faulconer and Williams, 1985.

37. Packer, 1985.

38. See, e.g., Chessick, 1990; Messer et al., 1988.

39. E.g., Chessick, 1990.

40. This may be a sign of a hidden linearity in some descriptions of hermeneutic ontology. Chessick (1986), for example, says that our "self-image . . . is determined by our developmental and life experiences" (p. 88), and "'man' has no essence but is constructed from birth by his specific culture" (p. 84). These passages lean heavily on the past and do not mention the equally important futurity of the person.

41. Shweder and Bourne, 1982.

42. Eastern thinkers, on the other hand, are likely to prefer a more contextual and sociocentric understanding of the self. Kojima (1984), for instance, has found that Japanese "do not think of themselves as exerting control over an environment that is utterly divorced from the self, nor over a self that stands apart from the environment" (p. 973). Instead, Japanese are more likely to identify the self with a particular temporal context, such as a work group.

43. Fuller, 1990, p. 52.

44. Gadamer, 1975, p. 240.

45. Emphasis added, Goffman, 1959, p. 133.

46. Dreyfus, 1979, ch. 7 and 8.

47. Sampson, 1985.

48. Slife, 1991b.

49. Frank, 1987.

50. Yalom, 1985.

51. Gadamer, 1975, p. 245.

52. Nichols, 1984, p. 431.

53. Polkinghorne, 1983, ch. 4.

BIBLIOGRAPHY

Abbot, V., Black, J. B., and Smith, E. E. (1985). The representation of scripts in memory. *Journal of Memory and Language, 24:*179–99.

Adler, A. (1974). The science of living. In W. S. Sahakian (ed.), *Psychology of Personality: Readings in theory.* Chicago: Rand McNally. (Original work published 1929)

Adler, A. (1958). *What Life Should Mean to You.* New York: Capricon Books.

Adler, A. (1969). *The Theory and Practice of Individual Psychology.* Patterson, N.J.: Littlefield.

Allport, F. H. (1962). A structuronomic conception of behavior: Individual and collective. *Journal of Abnormal and Social Psychology, 64:*3–30.

Anderson, J. (1980). *Cognitive Psychology and Its Implications.* New York: Freeman Press.

Arcaya, J. (1991). Making time for memory: A transcendental approach. *Theoretical and Philosophical Psychology, 11*(2):75–90.

Archer, E. J. (1960). Re-evaluation of the meaningfulness of all possible trigrams. *Psychological Monographs, 74,*(10) (Whole No. 497).

Aries, P. (1962). *Centuries of Childhood.* New York: Vintage Books.

Ariotti, P. E. (1975). The concept of time in Western antiquity. In J. T. Fraser and N. Lawrence (eds.), *The Study of Time II: 2nd Conference of the International Society for the Study of Time. Summer, 1973.* New York: Springer-Verlag.

Aristotle. (1990). Physics. In M. Adler (ed.), *Great Books of the Western World,* vol. 7. Chicago: Encyclopedia Britannica, pp. 257–355.

Arlow, J. A. (1989). Psychoanalysis. In R. J. Corsini and D. Wedding (eds.), *Current Psychotherapies,* 4th ed., pp. 19–64. Itasca, Ill.: F. E. Peacock.

Ashcraft, M. H. (1989). *Human Memory and Cognition.* New York: Scott, Foresman.

Aveni, A. F. (1989). *Empires of Time.* New York: Basic Books.

Baldwin, J. D., and Baldwin, J. I. (1973). The role of play in social organization. *Primaets, 14:*369–81.

Ballif, J. R., and Dibble, W. E. (1969). *Conceptual Physics.* New York: John Wiley and Sons.

Baltes, P. B. (1983). Life-span developmental psychology: Observations on history and theory revisited. In R. M. Lerner, (ed.), *Regressions in Mental Development: Basic Phenomena and Theories.* Hillsdale, N.J.: Lawrence Erlbaum.

Baltes, P. B., and Brim, O. G. (1983). *Life-Span Development and Behavior.* New York: Academic Press.

Baltes, P. B., and Brim, O. G. (1984). *Life-Span Development and Behavior.* New York: Academic Press.

Banaji, M. R., and Crowder, R. G. (1989). The bankruptcy of everyday memory. *American Psychologist, 44:*1185–93.

Bandura, A. (1973). *Aggression: A Social Learning Analysis.* Englewood Cliffs, N.J.: Prentice-Hall.

Bandura, A. (1986). *Social Foundations of Thought and Action: A Social Cognitive Theory.* Englewood Cliffs, N.J.: Prentice-Hall.

Barrow, I. (1916). *Geometrical Lectures,* trans. J. M. Child. London: Open Court Publishing Company. (Original work published 1735)

Bartlett, F. C. (1932). *Remembering: A Study in Experimental and Social Psychology.* London: Cambridge University Press.

Bateson, G. (1972). *Steps to an Ecology of Mind.* New York: Ballantine Books.

Bateson, G. (1978). *Mind and Nature.* New York: E. P. Dutton.

Bateson, G., Jackson, D. D., Haley, J., and Weakland, J. (1956). Towards a theory of schizophrenia. *Behavior Science, 1:*251–64.

Baumrind, D. (1972). Socialization and instrumental incompetence in young children. In W. W. Hartup (ed.), *The Young Child: Reviews of Research,* vol. 2. Washington, D.C.: National Association for the Education of Young Children.

Beck, A. J., and Weishaar, M. E. (1989). Cognitive therapy. In R. J. Corsini and D. Wedding (eds.), *Current Psychotherapies,* 4th ed. pp. 285–322. Itasca, Ill.: F. E. Peacock.

Becvar, D. S., and Becvar, R. J. (1988). *Family Therapy: A Systematic Integration.* Boston: Allyn and Bacon.

Belsky, J., and Rovine, M. (1988). Nonmaternal care in the first year of life and infant parent attachment security. *Child Development, 59:*157–67.

Benjamin, C. (1966). Ideas of time in the history of philosophy. In J. T. Fraser (ed.), *The Voices of Time,* pp. 3–30. New York: George Braziller.

Bertalanffy, L. von (1928). *Modern Theories of Development.* New York: Harper and Row.

Bertalanffy, L., von (1967). *Robots, Men, and Minds.* New York: George Braziller.

Bertalanffy, L., von (1968). *General System Theory.* New York: George Braziller.

Bertalanffy, L., von (1981). *A Systems View of Man.* Boulder, Colo.: Westview Press.

Bettelheim, Bruno. (1983). *Freud and Man's Soul.* New York: Alfred A. Knopf.

Binswanger, L. (1956). Existential analysis and psychotherapy. In E. Fromm-Reichman and J. L. Moreno (eds.), *Progress in Psychotherapy,* pp. 144–68. New York: Grune and Stratton.

Binswanger, L. (1958). The existential analysis school of thought. In R. May, E. Angel, H. Ellenberger (eds.), *Existence: A New Dimension in Psychiatry and Psychology.* New York: Basic Books.

Binswanger, L. (1967). *Being-in-the-World: Selected Papers of Ludwig Binswagner.* In J. Needleman (ed.), New York: Harper Torchbooks.

Birren, J., and Cunningham, W. (1985). Research on the psychology of aging: Principles, concepts, and theory. In J. Birren and K. W. Schaie (eds.), *Handbook of the Psychology of Aging,* 2nd ed. New York: Von Nostrand Reinhold.

Bjorklund, D. F. (1989). *Children's Thinking, Developmental Function and Individual Differences.* Pacific Grove, Calif.: Brooks-Cole.

Blanck, R., and Blanck, G. (1986). *Beyond Ego Psychology: Developmental Objects Relations Theory.* New York: Columbia University Press.

Bohm, D. (1980). *Wholeness and the Implicate Order.* London: Routledge and Kegan Paul.

Boss, M. (1963). *Psychoanalysis and Daseinanalysis.* New York: Basic Books.

Bower, G. H. (1986). Mood and Memory. *American Psychologist, 36:*129–48.

Brand, M. (1976). *The Nature of Causation.* Urbana: University of Illinois Press.

Bransford, J. D., and Franks, J. J. (1971). The abstraction of linguistic ideas. *Cognitive Psychology, 2:*331–50.

Bransford, J. D., and Franks, J. J. (1972). The abstraction of linguistic ideas: A review. *Cognition: International Journal of Cognitive Psychology, 1:*211–49.

Brim, O. G., and Kagan, J. (eds.). (1980). *Constancy and Change in Human Development.* Cambridge, Mass.: Harvard University Press.

Buhler, C. (1959). Theoretical observations about life's basic tendencies. *American Journal of Psychotherapy, 13:*561–81.

Bunge, M. (1959). *Causality.* Cambridge, Mass: Harvard University Press.

Bunge, M. (1963). *The Myth of Simplicity.* Englewood Cliffs, N.J.: Prentice-Hall.

Burtt, E. A. (1954). *The Metaphysical Foundations of Modern Physical Science.* Garden City, N.Y.: Doubleday.

Cameron, C., and Rychlak, J. F. (1985). *Personality Development and Psychopathology: A Dynamic Approach,* 2nd ed. Boston: Houghton Mifflin.

Campbell, D. J., and Stanley, J. C. (1963). *Experimental and Quasi-experimental Designs for Research.* Chicago: Rand McNally.

Capra, F. (1982). *The Turning Point: Science, Society, and the Rising Culture.* New York: Bantam Books.

Carlson, D. B., and LaBarba, R. C. (1979). Maternal emotionality during pregnancy and reproductive outcome. A review of the lierature. *International Journal of Behavioral Development, 2:*331–37.

Chessick, R. D. (1986). Heidegger for psychotherapists. *American Journal of Psychotherapy, 40*(1): 83–95.

Chessick, R. D. (1990). Hermeneutics for psychotherapists. *American Journal of Psychotherapy, 44*(2):256–73.

Chomsky, N. (1966). *Cartesian Linguistics.* New York: Harper and Row.

Chomsky, N. (1968). *Language and Mind.* New York: Harcourt Brace Jovanovich.

Clarke, A. M., and Clarke, A. D. (1977). *Early Experience: Myth and Evidence.* New York: The Free Press.

Corey, G. (1990). *Theory and Practice of Group Counseling,* 3rd ed. Pacific Grove, Calif.: Brooks-Cole.

Corsini, R., and Wedding, D. (1989). *Current Psychotherapies,* 4th ed. Itasca, Ill.: Peacock.

Coveney P., and Highfield, R. (1990). *The Arrow of Time.* New York: Fawcett Columbine.

Craig, G. J. (1989). *Human Development,* 5th ed. Englewood Cliffs, N.J.: Prentice-Hall.

Craik, F. I. M., and Lockhart, R. S. (1972). Levels of processing: A framework for memory research. *Journal of Verbal Learning and Verbal Behavior, 11*:671–84.

Craik, F. I. M., and Watkins, M. J. (1973). The role of rehearsal in short-term memory. *Journal of Verbal Learning and Verbal Behavior, 12*:599–607.

Crider, A. B., Goethals, G. R., Kavanaugh, R. D., and Solomon, P. R. (1989). *Psychology,* 3rd ed..Glenview, Ill.: Scott, Foresman.

Darwin, C. (1962). The *Origin of Species,* New York: Collier Books. (Original work published 1872)

DeBroglie, L. A. (1949). A general survey of the scientific work of Albert Einstein. In P. Schilpp (ed.), *Albert Einstein, Philosopher Scientist,* Vol. 1, pp. 38–51. New York: Harper and Row.

Dilthey, W. (1976). The rise of hermeneutics. In P. Connerton (ed.), *Critical Society,* pp. 104–16. New York: Penguin Books. (Original work published 1900)

Dollard, J., and Miller, N. E. (1950). *Personality and Psychotherapy: An Analysis in Terms of Learning, Thinking, and Culture.* New York: McGraw-Hill.

Dreyfus, H. S. (1979). *What Computers Can't Do.* New York: Harper and Row.

Dreyfus, H. L., and Dreyfus, S. E. (1986). *Mind Over Machine: The Power of Human Intuition and Expertise in the Era of the Computer.* New York: The Free Press.

Durkheim, E. (1966). *Suicide.* New York: Free Press. (Original work published 1897)

Einstein, A. (1990). Relativity: The special and general theory, trans. Robert W. Larson. In M. Adler (ed.), *Great Books of the Western World.* Chicago: Encyclopedia Britanica. (Original work published 1961)

Ellis, A. (1989). Rational emotive therapy. In R. J. Corsini and D. Wedding (eds.), *Current Psychotherapies,* 4th ed., pp. 197–240. Itasca, Ill.: F. E. Peacock.

Ellis, H. (1900). The analysis of the sexual impulse. *The Alienist and the Neurologist, 21*:247–62.

Ellis, H. C., and Hunt, R. R. (1989). *Fundamentals of Human Memory and Cognition,* 4th ed. Dubuque, Iowa: W. C. Brown.

Emde, R. N. (1981). Changing models of infancy and the nature of early development: Remodeling the foundation. *Journal of the American Psychoanalytic Association, 29*:179–219.

Emde, R. N., and Harmon, R. J. (1984). Entering a new era in the search for developmental continuities. In R. Emde and R. Harmon (eds.), *Continuities and Discontinuities in Development.* New York: Plenum Press.

Ermath, M. (1978) *Wilhelm Dilthey: The Critique of Historical Reason.* Chicago: University of Chicago Press.

Evans, R. I. (1973). *Jean Piaget: The Man and His Ideas.* New York: E. P. Dutton.

Ewen, R. B. (1988). *An Introduction to Theories of Personality,* 3rd ed. Hillsdale, N.J.: Lawrence Erlbaum Associates.

Faulconer, J. (1990). Heidegger and psychological explanation: Taking account of Derrida. In J. Faulconer and R. Williams (eds.), *Reconsidering Psychology: Perspectives from Continental Philosophy,* pp. 116–35. Pittsburgh: Duquesne University Press.

Faulconer, J., and Williams, R. (1985). Temporality in human action: An alternative to positivism and historicism. *American Psychologist, 40*:1179–88.

Faulconer, J., and Williams, R. (1990). Reconsidering psychology. In J. Faulconer and R. Williams (eds.), *Reconsidering Psychology: Perspectives from Continental Philosophy,* pp. 9–60. Pittsburgh: Duquesne University Press.

Fischer, K. W. (1984). Detecting developmental discontinuities: Methods and measurement. In R. N. Emde and R. J. Harmon (eds.), *Continuities and Discontinuities in Development.* New York: Plenum Press.

Forsyth, D. R. (1990). *Group Dynamics,* 2nd ed. Pacific Grove, Calif.: Brooks-Cole.

Frank, J. D. (1987). Psychotherapy, rhetoric, and hermeneutics: Implications for practice and research. *Psychotherapy, 24*:293–301.

Freud, S. (1933). The anatomy of mental personality. (Lecture 31), In J. H. Sprott (trans.), *New Introductory Lectures on Psycho-Analysis.* New York: W. W. Norton & Company, Inc.

Freud, S. (1957a). Five lectures on psycho-analysis, In J. Strachey (ed. and trans.), *The Standard Edition of the Complete Psychological Works of Sigmund Freud,* (vol. 11. pp. 9–56). London: Hogarth Press. (Original work published 1909)

Freud, S. (1957b). Papers on metapsychology. In J. Strachey (ed. and trans.), *The Standard Edition of the Complete Psychological Works of Sigmund Freud,* (vol. 14. pp. 105–215). London: Hogarth Press. (Original work published 1915)

Freud, S. (1959a). Letter 52, In J. Strachey (ed. and trans.), *The Standard Edition of the Complete Psychological Works of Sigmund Freud,* (vol. 7. pp. 223–239). London: Hogarth Press. (Original work published 1897)

Freud, S. (1959b). Inhibitions, symptoms, and anxiety. In J. Strachey (ed. and trans.), *The Standard Edition of the Complete Psychological Works of Sigmund Freud,* (vol. 20, pp. 77–175). London: Hogarth Press. (Original work published 1926)

Freud, S. (1961). The ego and the id. In J. Strachey (ed. and trans.), *The Standard Edition of the Complete Psychological Works of Sigmund Freud,* (vol. 19. pp. 12–66). London: Hogarth Press. (Original work published 1923)

Freud, S. (1963). Introductory lectures on psycho-analysis (parts I and II). In J. Strachey (ed. and trans.), *The Standard Edition of the Complete Psychological Works of Sigmund Freud,* (vol. 15. pp. 15–239). London: Hogarth Press. (Original work published 1916)

Freud, S. (1989). The ego and the id. In P. Gay (ed.), *The Freud Reader.* New York: W. W. Norton. (Original work published 1923)

Freud, S. (1990). Beyond the pleasure principle. In M. Adler (ed.), *Great Books of the Western World,* trans. C. Hubback. Chicago: Encyclopaedia Britanica. (Original work published 1920)

Froberg, W., and Slife, B. D. (1987). Overcoming obstacles to the implementation of Yalom's model of inpatient group psychotherapy. *International Journal of Group Psychotherapy, 37*(3):371–88.

Fuller, A. R. (1990). *Insight into Value: An Exploration of the Premises of a Phenomenological Psychology.* Albany, N.Y.: SUNY Press.

Gadamer, H. G. (1975). *Truth and Method,* trans. G. Barden and J. Cumming. New York: Seabury Press.

Garfield, S., and Kurz, R. (1976). Clinical psychologists in the 1970s. *American Psychologist, 31*:1–9.

Garson, J. W. (1971). Here and now. In E. Freeman, and W. Sellars (eds.), *Basic Issues in the Philosophy of Time.* LaSalle, Ill: Open Court.

Gergen, K. J. (1985). The social constructionist movement in modern psychology. *American Psychologist, 40*:266–75.

Goffman, E. (1959). The moral career of the mental patient. *Psychiatry, 22*:123–42.

Goldenberg, I., and Goldenberg, H. (1991). *Family Therapy: An Overview,* 3rd ed. Belmont, Calif.: Brooks-Cole.

Goldstein, K. (1939). *The Organism.* New York: American Book Company.

Haley, J. (1959). The family of the schizophrenic. *American Journal of Nervous and Mental Diseases, 129*:357–74.

Haley, J. (1976). *Problem-Solving Therapy.* San Francisco: Jossey-Bass.

Hall, C. S., and Lindzey, G. (1978). *Theories of Personality,* 3rd ed. New York: John Wiley and Sons.

Hall, E. (1982). *Child Psychology Today.* New York: Random House.

Harris, E. E. (1988). *The Reality of Time.* Albany, N.Y.: SUNY Press.

Hawking, S. (1988). *A Brief History of Time,* New York: Bantam Books.

Heidegger, M. (1962). *Being and Time,* trans. J. Macquarrie and E. S. Robinson. New York: Harper and Row.

Heisenberg, W. (1958). *Physics and Philosophy: The Revolution in Modern Science.* New York: Harper & Brothers Publishers.

Hempel, C. G. (1965). *Aspects of Scientific Explanation and Other Essays in the Philosophy of Science.* New York: The Free Press.

Holton, G. (1973). *Thematic Origins of Scientific Thought: Kepler to Einstein.* Cambridge, Mass.: Harvard University Press.

Hume, D. (1911). *A Treatise of Human Nature.* New York: Dutton. (Original work published 1739–1740)

Jacobson, N., and Margolin, G. (1979). *Marital Therapy: Strategies Based on Social Learning and Behavioral Exchange Principles.* New York: Brunner/Mazel.

Jenson, L. C., and Kingston, M. (1986). *Parenting.* New York: CBS College Publishing.

Jones, W. T. (1969). Kant to Wittgenstein and Sartre, 2nd ed., vol 4 of *A history of Western philosophy*. New York: Harcourt, Brace, and World.

Jung, C. G. (1954). The practice of psychotherapy. In H. Read, M. Fordham, and G. Alder (eds.), *The Collected Works of C. G. Jung*, (vol. 16). New York: Pantheon Books.

Jung, C. G. (1959). The archetypes and the collective unconscious. In H. Read, M. Fordham, and G. Alder (eds.), *The Collected Works of C. G. Jung*, (vol. 9). New York: Pantheon Books.

Jung, C. G. (1960a). The structure and dynamics of the psyche. In H. Read, M. Fordham, and G. Adler (eds.), *The Collected Works of C. G. Jung*, (vol. 8). New York: Pantheon Books.

Jung, C. G. (1960b). The psychogenesis of mental disease. In H. Read, M. Fordham, and G. Adler (eds.), *The Collected Works of C. G. Jung*, (vol. 3). New York: Pantheon Books.

Jung, C. G. (1961). Freud and psychoanalysis. In H. Read, M. Fordham, and G. Adler (eds.) *The Collected Works of C. G. Jung*, (vol. 4). New York: Pantheon Books.

Kagan, J. (1971). *Change and Continuity in Infancy*. New York: John Wiley and Sons.

Kagan, J. (1980). Perspectives on continuity. In O. G. Brim, Jr., and J. Kagan (eds.), *Constancy and Change in Human Development*, pp. 26–74. Cambridge, Mass.: Harvard University Press.

Kagan, J. (1981). *The Second Year: The Emergence of Self-Awareness*. Cambridge, Mass.: Harvard University Press.

Kagan, J. (1984). Continuity and change in the opening years of life. In R. N. Emde and R. J. Harmon (eds.), *Continuities and Discontinuities in Development*, pp. 15–40. New York: Plenum Press.

Kagan, J., Kearsley, R., and Zelazo, P. (1978). *Infancy: Its Place in Human Development*. Cambridge, Mass.: Harvard University Press.

Kagan, J., Lapidus, D., and Moore, M. (1978). Infant antecedants of cognitive functioning. *Child Development, 49*:1005–23.

Kagan, J., and Moss, H. A. (1962). *Birth to Maturity*. New York: John Wiley and Sons.

Kant, I. (1966). *Critique of Pure Reason*, trans. F. Max Muller. New York: Doubleday Anchor Books. (Original work published 1781)

Kant, I. (1990). *Critique of Pure Reason,* trans. J. M. D. Meiklejohn. Chicago: University of Chicago Press. (Original work published 1781)

Kaufman, Y. (1989). Analytical psychotherapy. In R. J. Corsini and D. Wedding (eds.), *Current Psychotherapies,* 4th ed., pp. 119–54. Itasca, Ill.: F. E. Peacock.

Keeney, B. P. (1983). *Aesthetics of Change.* New York: Guilford.

Kempler, W. (1965). Experiential family therapy. *International Journal of Group Psychotherapy, 15:*57–71.

Kempler, W. (1968). Experiential psychotherapy with families. *Family Process, 7:*88–99.

Keuhnel, J. M., and Liberman, R. P. (1986). Behavior modification. In I. L. Kutash and A. Wolf (eds.), *Psychotherapist's Casebook,* pp. 240–62. San Francisco: Jossey-Bass.

Kintsch, W. (1977). *Memory and Cognition,* 2nd ed. New York: John Wiley and Sons.

Klaus, M. H., and Kennel, J. H. (1976). *Maternal-Infant Bonding: The Impact of Early Separation or Loss on Family Development.* St. Louis: C. V. Mosby.

Klaus, M. H., and Kennel, J. H. (1982) *Parent-Infant Bonding.* St. Louis: C. V. Mosby.

Kockelmans, J. J. (1965). *Martin Heidegger.* Pittsburgh: Duquesne University Press.

Kohler, W. (1927). *The Mentality of Apes.* New York: Harcourt Brace.

Kohut, H. (1971). *The Analysis of Self.* New York: International Universities Press.

Kohut, H. (1978). *The Psychology of the Self,* ed. A. Goldberg. New York: International Universities Press.

Kojima, H. (1984). A significant stride toward the comparative study of control. *American Psychologist, 39:*972–73.

Kuhn, T. S. (1970). *The Structure of Scientific Revolutions,* 2nd ed. Chicago: University of Chicago Press.

Kuiken, D. (1991). *Mood and Memory.* Newbury Park, Calif. Sage Publications Inc.

Lakin, M. (1985). *The Helping Group: Therapeutic Principles and Issues.* Reading, Mass.: Addison-Wesley Publishing.

Lamb, M. E. (1982a). Early contact and maternal-infant bonding: One decade later. *Pediatrics, 70:*763–68.

Lamb, M. E. (1982b). Parental behavior and child development in nontraditional families. An introduction. In M. E. Lamb (ed.), *Nontraditional Families.* Hillsdale, N.J.: Lawrence Erlbaum.

Leahey, T. H. (1987). *A History of Psychology: Main Currents in Psychological Thought.* Englewood Cliffs, N.J.: Prentice-Hall.

Lee, D. (1950). Lineal and nonlineal codifications of reality. *Psychosomatic Medicine, 12:*89–97.

Lewin, K. (1936). *Principles of Topological Psychology.* New York: McGraw-Hill.

Lewis, V. E., and Williams, R. N. (1989). Mood-congruent versus mood-state–dependent learning: Implications for a view of emotion. *Journal of Social Behavior and Personality, 4*(2):157–71.

Lipsitt, L. P. (1983). Stress in infancy: Toward understanding the origins of coping behavior. In N. Garmezy and M. Rutter (eds.), *Stress, Coping, and Development in Children.* New York: McGraw-Hill.

Locke, J. (1990). *An Essay Concerning Human Understanding.* Chicago: The University of Chicago Press. (Original work published 1689)

Locke, N. (1961). *Group Psychoanalysis: Theory and Practice.* New York: New York University Press.

Loftus, E. (1979). *Eyewitness Testimony.* Cambridge, Mass: Harvard University Press.

Lovejoy, A. O. (1961). *The Reason, the Understanding, and Time.* Baltimore: Johns Hopkins Press.

Mach, E., (1959). *The Analysis of Sensations.* New York: Dover Publications.

MacKinnon, L. (1983). Contrasting strategic and Milan therapies. *Family Process, 22:*425–40.

Maddi, S. R. (1989). *Personality Theories: A Comparative Analysis,* 5th ed. Chicago: Dorsey.

Mahler, M. S. (1968). *On Human Symbiosis and the Vicissitudes of Individuation.* New York: International Universities Press.

Main, M. (1973). *Exploration, play, and cognitive functioning as related to mother-child attachment.* Unpublished dissertation, John Hopkins University.

Martindale, C. (1981). *Cognition and Consciousness.* Homewood, Ill.:
    Dorsey.

Matas, L., Arent, R., and Sroufe, L. (1978). Continuity in adaptation in the
    second year: The relationships between quality of attachment and
    later competence. *Child Development, 49:*547–56.

Matlin, M. (1983). *Cognition.* New York: Holt, Rinehart, and Winston.

Maturana, H. (1978). Biology of language: The epistemology of reality. In
    G. A. Miller and E. Lennerberg (eds.), *Psychology and Biology of
    Language and Thought: Essays in Honor of Eric Lennerberg,* pp.
    27–63. New York: Academic Press.

May, R., Angel, E., and Ellenberger, H. (1958). *Existence: A New Dimension
    in Psychiatry and Psychology.* New York: Basic Books.

May, R., and Yalom, I. (1989). Existential psychotherapy. In R. J. Corsini
    and D. Wedding (eds.), *Current Psychotherapies,* 4th ed. pp. 363–
    404. Itasca, Ill.: F. E. Peacock.

McCall, R. B. (1979). The development of intellectual functioning in infancy
    and the prediction of later IQ. In J. D. Osofsky (ed.), *Handbook of
    Infant Development.* New York: John Wiley and Sons.

McGrath, J. E., and Kelly, J. R. (1986). *Time and Human Interaction:
    Toward a Social Psychology of Time.* New York: Guilford.

Meichenbaum, D. (1986). Cognition behavior modification. In F. H. Kanfer
    and A. P. Goldstein (eds.), *Helping People Change: A Textbook of
    Methods,* 3rd ed. New York: Pergamon.

Melkman, R. (1988). *The Construction of Objectivity: A New Look at the
    First Months of Life.* New York: Karger.

Mendelson, E. (1980). The continuous and the discrete in the history of
    science. In O. G. Brim, Jr., and J. Kagan (eds.), *Constancy and
    Change in Human Development,* pp. 75–112. Cambridge, Mass.:
    Harvard University Press.

Menninger, K. (1966). *The Crime of Punishment.* New York: Viking Press.

Merleau-Ponty, M. (1962). *Phenomenology of Perception.* London:
    Routledge and Kegan Paul.

Messer, S., Sass, L., and Woolfolk, R. (eds.) (1988). *Hermeneutics
    and Psychological Theory: Interpretative Perspectives on Personal-
    ity, Psychotherapy, and Psychopathology.* New Brunswick, N.J.:
    Rutgers University Press.

Miller, N. E., and Dollard, J. (1941). *Social Learning and Imitation.* New
    Haven, Conn.: Yale University Press.

Minsky, M. (1986). *The Society of Mind.* New York: Simon and Schuster.

Minuchin, S. (1974). *Families and Family Therapy.* Cambridge, Mass.: Harvard University Press.

Minuchin, S., and Fishman, L. (1981). *Family Therapy Techniques.* Cambridge, Mass.: Harvard University Press.

Morris, C. G. (1989). *Psychology,* 7th ed. Englewood Cliffs, N.J.: Prentice-Hall.

Morris, R. (1984). *Time's Arrows.* New York: Simon and Schuster.

Moss, H. A., and Susman, E. J. (1980). Longitudinal study of personality development. In O. G. Brim, Jr., and J. Kagan (eds.), *Constancy and Change in Human Development,* pp. 530–95. Cambridge, Mass.: Harvard University Press.

Mueller-Vollmer, K. (1990). *The Hermeneutics Reader.* New York: Continuum.

Mullan, H. (1979). An existential group psychotherapy. *International Journal of Group Psychotherapy, 29:*163–74.

Mumford, L., (1934). *Techniques and Civilization.* London: Rutledge.

Neisser, U. (1978). Memory: What are the important questions? In Gruenberg, M. M., Morris, P. E. and Sykes, R. N. (eds.). *Practical Aspects of Memory.* New York: Academic Press.

Neisser, U. (1991). A case of misplaced nostalgia. *American Psychologist, 46:*34–36.

Neugarten, B. L. (1980). Must everything be a mid-life crisis? *Annual Editions, Human Development 80/81,* pp. 289–90. Guilford, Conn.: Dushkin.

Newell, A., and Simon, H. A. (1972). *Human Problem Solving.* Englewood Cliffs, N.J.: Prentice-Hall.

Newton, I. (1990). *Mathematical Principles of Natural Philosophy,* trans. A. Motte (revised by Florian Cajori). Chicago: University of Chicago Press. (Original work published 1687)

Newton, I. (1990). *Optics.* Chicago: University of Chicago Press. (Original work published 1704)

Nichols, M. P. (1984). *Family Therapy.* New York: Gardner Press.

Nicolson, I. (1980). Mutable time. In J. Grant, and C. Wilson (eds.) *The Book of Time.* North Pomfret, Vt.: David and Charles.

Noble, C. E. (1952). The role of stimulus meaning (m) in serial verbal learning. *Journal of Experimental Psychology, 43:*437–46.

Oettinger, A. G. (1979). Preface. In H. Dreyfus, *What Computers Can't Do.* New York: Harper and Row.

Oppenheim, R. W. (1981). Ontogenetic adaptations and retrogressive processes in the development of the nervous system and behavior: A neuroembryological perspective. In K. J. Connolly and H. F. R. Prechtel (eds.), *Maturation and Development: Biological and Psychological Perspectives.* Philadelphia: J. B. Lippincott.

Ornstein, R. E. (1972). *The Psychology of Consciousness.* New York: Viking.

Overton, W. F., and Reese, H. W. (1973). Models of development: Methodological implications. In J. R. Nesselroade and H. W. Reese (eds.), *Life-Span Developmental Psychology: Methodological Issues.* New York: Academic Press.

Packer, M. J. (1985). Hermeneutic inquiry in the study of human conduct. *American Psychologist, 40:*1081–93.

Pepper, S. C. (1970). *World Hypotheses.* Berkeley: University of California Press.

Perls, F. (1973). *The Gestalt Approach and Eyewitness to Therapy.* New York: Bantam Books.

Perrig, W., and Kintsch, W. (1985). Propositional and situational representations of text. *Journal of Memory and Language, 24:*503–18.

Phares, E. J. (1984). *Introduction to Personality.* Glenview, Ill.: Scott, Foresman.

Piaget, J. (1967). *Six Psychological Studies.* New York: Random House.

Piaget, J. (1970). Piaget's theory. In P. H. Mussen (ed.), *Manual of Child Psychology,* 3rd ed., vol. 1, pp. 703–32. New York: John Wiley and Sons.

Piaget, J. (1973). *The Child and Reality: Problems of Genetic Psychology.* New York: Viking.

Pittendrigh, C. S. (1972). On temporal organization in living systems. In H. Yaker, H. Osmond, and F. Cheek (eds.), *The Future of Time,* pp. 179–218. New York: Anchor Books.

Plomin, R. (1983). Childhood temperament. In B. Lahey and A. Kazdin (eds.), *Advances in Clinical Child Psychology, 6:*1–78.

Polkinghorne, D. (1983). *Methodology for the Human Sciences.* Albany, N.Y.: SUNY Press.

Porter, R. (1980). The history of time. In J. Grant, and C. Wilson (eds.) *The Book of Time.* North Pomfret, Vt.: David and Charles.

Prigogine, I., and Stengers, I. (1984). *Order Out of Chaos.* London: Heinemann.

Rakover, S. S. (1990). *Metapsychology: Missing Links in Behavior, Mind and Science.* New York: Paragon.

Reeves, R. (1982). *American Journey.* New York: Simon and Schuster.

Reiser, B. J., Black, J. B., and Abelson, R. P. (1985). Knowledge structures in the organization and retrieval of autobiographical memories. *Cognitive Psychology, 17:*89–137.

Reynolds, A. G., and Flag, P. W. (1977). *Cognitive Psychology.* Cambridge, Mass.: Winthrop Publishers.

Ricoeur, P. (1985). *Time and Narrative.* Chicago: University of Chicago Press.

Rickman, H. P. (1979). *Wilhelm Dilthey: Pioneer of the Human Studies.* Berkley: University of California Press.

Robbins, L. (1978). Sturdy childhood predictors of adult antisocial behavior: Replications from longitudinal studies. *Psychological Medicine, 8:*611–22.

Robinson, D. N. (1976). *An Intellectual History of Psychology.* New York: Macmillon.

Rogers, C. (1963). Learning to be free. In S. Farber and R. Wilson (eds.), *Conflict and Creativity: Control of the Mind.* (Part 2) pp. 268–88. New York: McGraw-Hill.

Rogers, C. (1970). *On Encounter Groups.* New York: Harper and Row.

Rose, F. (1985). The Black Knight of AI. *Science 85, 6:*47–51.

Rose, S. D. (ed.). (1980). *A Casebook in Group Therapy: A Behavioral-Cognitive Approach.* Englewood Cliffs, N.J.: Prentice-Hall.

Ross, A. O. (1987). *Personality: The Scientific Study of Complex Human Behavior.* New York: Holt, Rinehart, and Winston.

Rudestam, K. E. (1982). *Experiential Groups in Theory and Practice.* Belmont, Calif.: Brooks/Cole.

Russell, B. (1971). Descriptions. In J. F. Rosenberg and C. Travis (eds.). *Reading in the Philosophy of Language.* Englewood Cliffs, N.J.: Prentice-Hall. (Original work published 1919)

Russell, B. (1959). *Wisdom of the West.* Garden City, N.Y.: Doubleday.

Rutter, M. (1984). Continuities and discontinuities in socio-emotional development: Empirical and conceptual perspectives. In R. N. Emde and R. J. Harmon (eds.), *Continuities and Discontinuities in Development,* pp. 41–63. New York: Plenum Press.

Rutter, M. (1989). The role of cognition in child development and disorder. In S. Chess, A. Thomas, and M. Hertzig (eds.), *Annual Progress in Child Psychiatry and Child Development,* pp. 77–101. New York: Brunner/Mazel.

Rutter, M., and Giller, H. (1983). *Juvenile Delinquency: Trends and Perspectives.* Harmondsworth, U.K.: Penguin.

Rutter, M., Quinton, D., and Liddle, C. (1983). Parenting in two generations: Looking backwards and looking forwards. In N. Madge (ed.), *Families at Risk.* London: Heinemann.

Rychlak, J. F. (1977). *The Psychology of Rigorous Humanism.* New York: John Wiley and Sons.

Rychlak, J. F. (1979). *Discovering Free Will and Personal Responsibility.* New York: Oxford Press.

Rychlak, J. F. (1981). *Introduction to Personality and Psychotherapy: A Theory-Construction Approach.* Boston: Houghton Mifflin.

Rychlak, J. F. (1988). *The Psychology of Rigorous Humanism,* 2nd ed. New York: New York University Press.

Rychlak, J. F., and Slife, B. D. (1984). Affection as a cognitive judgmental process: A theoretical assumption put to test through brain-lateralization methodology. *Journal of Mind and Behavior,* 5:131–50.

Ryckman, R. M. (1989). *Theories of Personality,* 4th ed. Belmont, Calif.: Wadsworth.

Sahakian, W. S. (ed.). (1974). *Psychology of Personality: Readings in Theory.* Chicago: Rand McNally.

Samelson, F. (1981). Struggle for scientific authority: The reception of Watson's behaviorism, 1913–1920. *Journal of the History of the Behavioral Sciences, 17*:399–425.

Sampson, E. E. (1985). The decentralization of identity: Toward a revised concept of personal and social order. *American Psychologist, 40*:1203–11.

Santrock, J.. W. (1986). *Life-Span Development,* 2nd ed. Dubuque, Iowa: W. C. Brown.

Sartre, J. P. (1947). *Existentialism.* New York: Philosophical Library.

Saussure, F. (1966). *Course in General Linguistics,* ed. C. Bally and A. Sechehaye, trans. W. Baskin. New York: McGraw-Hill. (Original work published 1907)

Scarr, S. (1987). Constructing psychology: Making facts and fables for our times. In S. Chess and A. Thomas (eds.), *Annual Progress in Child*

Psychiatry and Child Development, pp. 43–68. New York: Brunner/Mazel.

Schaffer, H. (1988). Child psychology: The future. In S. Chess, A. Thomas, and M. Hertzig (eds.), *Annual Progress in Child Psychiatry and Child Development,* pp. 89–112. New York: Brunner/Mazel.

Schaie, K. W., and Hertzog, C. (1985). Measurement in the psychology of adulthood and aging. In J. Birren and K. W. Schaie (eds.), *Handbook of the Psychology of Aging,* 2nd ed. New York: Von Nostrand Reinhold.

Schank, R. C., and Abelson, R. P. (1977). *Scripts, Plans, Goals, and Understanding.* Hillsdale, N.J.: Erlbaum.

Schiamberg, L. B., and Smith, K. U. (1982). *Human Development.* New York: Macmillan.

Schrag, C. O. (1990). Explanation and understanding in the science of human behavior. In J. Faulconer and R. Williams (eds.) *Reconsidering Psychology: Perspectives from Continental Philosophy.* Pittsburgh: Duquesne University Press.

Searle, J. (1985, August). *The Prospects for Cognitive Sciences.* Paper presented at the meeting of the American Psychological Association, Los Angeles, Calif.

Selvini-Palazoli, M., Boscolo, L., Cecchin, G., and Prata, G. (1978). *Paradox and Counter Paradox.* New York: Jason Aronson.

Shweder, R. A., and Bourne, E. (1982). Does the concept of the person vary cross-culturally? In A. J. Marsella and G. White (eds.), *Cultural Concepts of Mental Health and Therapy,* pp. 97–137. Boston: Reidel.

Simon, H. A. (1979). *Models of Thought.* New Haven, Conn.: Yale University Press.

Singer, J. L., and Kolligan, J. (1987). Personality: Developments in the study of private experience. *Annual Review of Psychology, 38*:533–74.

Skinner, B. F. (1957). *Verbal Behavior.* New York: Appleton-Century-Crofts.

Skinner, B. F. (1974). *About Behaviorism.* New York: Knopf.

Slife, B. D. (1981). Psychology's reliance on linear time: A reformulation. *Journal of Mind and Behavior, 1*:27–46.

Slife, B. D. (1987a). Telic and mechanistic explanations of mind and meaningfulness: An empirical illustration. *Journal of Personality, 55*:445–66.

Slife, B. D. (1987b). The perils of eclecticism as therapeutic orientation. *Theoretical and Philosophical Psychology, 7*:94–103.

Slife, B. D. (1988, August). *Linear Time Assumptions in Family and Group Psychotherapy: Inadequacies and Alternatives.* Paper presented at the meeting of the American Psychological Association, Atlanta, Ga.

Slife, B. D. (1989, August). *The Role of Time in Personality Explanation.* Paper presented at the meeting of the American Psychological Association, August, New Orleans, La.

Slife, B. D. (1991a). *Individualism, Intimacy, and Group Psychotherapy.* Paper presented at the meeting of the American Psychological Association, San Francisco, Calif.

Slife, B. D. (1991b). *The Influence of Temporal Assumptions on Conceptions of Self.* Paper presented at the meeting of the American Psychological Association, San Francisco, Calif.

Slife, B. D. (1991c). *Psychology's Paradigm: Linear Time.* Paper presented at the meeting of the American Psychological Association, San Francisco, Calif.

Slife, B. D. (in press). Newtonian time and psychological explanation. *Journal of Mind and Behavior.*

Slife, B. D., and Barnard, S. (1988). Existential and cognitive psychology: Contrasting views of consciousness. *Journal of Humanistic Psychology, 28,* (3):119–36.

Slife, B. D., and Lanyon, J. (1991). Accounting for the power of the here-and-now: A theoretical revolution. *International Journal of Group Psychotherapy, 41*(2):145–67.

Slife, B. D., Miura, S., Thompson, L. W., Shapiro, J. L., and Gallagher, D. (1984). Differential recall as a function of mood disorder in clinically depressed patients: Between- and within-subject differences. *Journal of Abnormal Psychology, 93*:391–400.

Slife, B. D., and Rychlak, J. F. (1981). Affection as a separate dimension of meaningfulness. *Contemporary Educational Psychology, 26*:337–48.

Slife, B. D., and Rychlak, J. F. (1982). The role of affective assessment in modeling aggressive behaviors. *Journal of Personality and Social Psychology, 43*:861–68.

Smith, P. B. (ed.). (1980). *Small Groups and Personality Change.* New York: Methuen.

Steiner, I. D. (1974). Whatever happened to the group in social psychology? *Journal of Experimental Social Psychology, 10*:94–108.

Stott, D. H. (1971). The child's hazards in utero. In J. G. Howells (ed.), *Modern Perspectives in International Child Psychiatry.* New York: Brunner/Mazel.

Thompson, R. A., and Lamb, M. E. (1984). Continuity and change in socioemotional development during the second year. In R. N. Emde and R. J. Harmon (eds.), *Continuities and Discontinuities in Development,* pp. 293–314. New York: Plenum Press.

Tolman, E. C. (1967). *Purposive Behavior in Animals and Men.* New York: Appleton-Century-Crofts.

Tomlinson-Keasey, C. (1985). *Child Development: Psychological, Sociocultural, and Biological Factors.* Homewood, Ill.: Dorsey Press.

Tomm, K. (1984). One perspective on the Milan approach: Part I, Overview of development, theory, and practice. *Journal of Marital and Family Therapy, 10:*113–25.

Tuttman, S. (1986). Theoretical and technical elements which characterize the American approaches to psychoanalytic group psychotherapy. *International Journal of Group Psychotherapy, 36:*499–515.

Van Inwagen, P. (1983). *An Essay on Free Will.* Oxford: Clarendon Press.

Vander Kolk, C. J. (1985). *Introduction to Group Counseling and Psychotherapy.*Columbus, Ohio: Merrill Publishing.

Varela, F. J., and Johnson, D. (1976). On observing natural systems. *CoEvolution Quarterly,* Summer, 26–31.

Von Neumann, J. (1951). The general and logical theory of automata. In L. Jeffress (ed.), *Cerebral Mechanisms in Behavior: The Hixon Symposium.* New York: John Wiley and Sons.

Waters, E., Wippman, J., and Sroufe, L. (1979). Attachment, positive affect, and competence in the peer group: Two studies in construct validation. *Child Development, 50:*821–29.

Watzlawick, P., Beavin, J., and Jackson, D. (1967). *Pragmatics of Human Communication.* New York: W. W. Norton.

Watzlawick, P., and Weakland, J. H. (1974). *The Interactional View: Studies at the Mental Institute, Palo Alto, 1965–74.* New York: W. W. Norton.

Watzlawick, P., Weakland, J. H., and Fisch, R. (1974). *Change: Principles of Problem Formation and Problem Resolution.* New York: W. W. Norton.

Weizenbaum, J. (1976). *Computer Power and Human Reason.* San Francisco: Freeman and Company.

Wessells, M. G. (1982). *Cognitive Psychology.* New York: Harper and Row.

Westcott, M. R. (1987). Minds, machines, models and metaphors: A commentary. *Journal of Mind and Behavior, 8:*281–90.

Whitaker, C. A. (1976). The hindrance of theory in clinical work. In P. J. Guerin (ed.), *Family Therapy: Theory and Practice,* pp. 154–64. New York: Gardner Press.

Whitrow, G. J. (1980). *The Natural Philosophy of Time,* 2nd ed. New York: Oxford University Press

Widdershoven, G. (1992). Hermeneutics and relativism: Wittgenstein, Gadamer, Habermas. *Theoretical and Philosophical Psychology, 12:*1–11.

Wiener, N. (1948). *Cybernetics, or Control and Communication in the Animal and the Machine.* Cambridge, Mass.: M.I.T. Press.

Williams, J. H. (1977). *Psychology of Women: Behavior in a Biosocial Context.* New York: W. W. Norton.

Williams, R. N. (1990a). Aristotle, time and temporality. *Theoretical and Philosophical Psychology, 10:*13–21.

Williams, R. N. (1990b). The metaphysics of things and discourse about them. In J. Faulconer and R. Williams (eds.), *Reconsidering Psychology: Perspectives from Continental Philosophy,* pp. 136–50. Pittsburgh: Duquesne University Press.

Wilson, G. T. (1989). Behavior therapy, In R. J. Corsini and D. Wedding (eds.), *Current Psychotherapies,* 4th ed., pp. 241–84. Itasca, Ill.: F. E. Peacock.

Winick, M. A., Meyer, K. K., and Harris, R. C. (1975). Malnutrition and environmental enrichment by early adoption. *Science, 190:*1173–75.

Winograd, T., and Flores, F. (1987). *Understanding Computers and Cognition: A New Foundation for Design.* Reading, Mass.: Addison-Wesley.

Wolf, A. (1963). The psychoanalysis of groups. In M. Rosenbaum and M. Berger (eds.), *Group Psychotherapy and Group Function.* New York: Basic Books.

Wolf, A. (1983). Psychoanalysis in groups. In H. I. Kaplan and B. J. Saddock (eds.), *Comprehensive Group Psychotherapy,* 2nd ed. Baltimore: Williams and Wilkins.

Wolf, A., and Kutash, I. L. (1986). Psychotherapy in groups. In A. Wolf and I. L. Kutash (eds.), *Psychotherapist's Casebook,* pp. 332–52. San Francisco: Jossey-Bass.

Wolf, F. A. (1981). *Taking the Quantum Leap.* San Francisco: Harper and Row.

Wolf, K. (ed.). (1950). *The Sociology of Georg Simmel.* New York: Collier-Macmillan.

Yalom, I. (1975). *Theory and Practice of Group Psychotherapy,* 2nd ed. New York: Basic Books.

Yalom, I. (1980). *Existential Psychotherapy.* New York: Basic Books.

Yalom, I. D. (1983). *Inpatient Group Psychotherapy.* New York: Basic Books.

Yalom, I. D. (1985). *Theory and Practice of Group Psychotherapy,* 3rd ed. New York: Basic Books.

Zinker, J. (1978). *Creative Process in Gestalt Therapy.* New York: Random House.

INDEX